The Theoretic Life - A Classical Ideal
and its Modern Fate

Alexander S. Rosenthal-Pubul

The Theoretic Life - A Classical Ideal and its Modern Fate

Reflections on the Liberal Arts

Alexander S. Rosenthal-Pubul
Center for Advanced Governmental Studies
Johns Hopkins University
Washington, DC, USA

ISBN 978-3-030-02280-8   ISBN 978-3-030-02281-5 (eBook)
https://doi.org/10.1007/978-3-030-02281-5

Library of Congress Control Number: 2018959844

© Springer Nature Switzerland AG 2018

This work is subject to copyright. All rights are reserved by the Publisher, whether the whole or part of the material is concerned, specifically the rights of translation, reprinting, reuse of illustrations, recitation, broadcasting, reproduction on microfilms or in any other physical way, and transmission or information storage and retrieval, electronic adaptation, computer software, or by similar or dissimilar methodology now known or hereafter developed.

The use of general descriptive names, registered names, trademarks, service marks, etc. in this publication does not imply, even in the absence of a specific statement, that such names are exempt from the relevant protective laws and regulations and therefore free for general use.

The publisher, the authors, and the editors are safe to assume that the advice and information in this book are believed to be true and accurate at the date of publication. Neither the publisher nor the authors or the editors give a warranty, express or implied, with respect to the material contained herein or for any errors or omissions that may have been made. The publisher remains neutral with regard to jurisdictional claims in published maps and institutional affiliations.

Cover image text: © Acropolis Museum NMA 5503, photo: Socratis Mavrommatis.

This Springer imprint is published by the registered company Springer Nature Switzerland AG
The registered company address is: Gewerbestrasse 11, 6330 Cham, Switzerland

Fig. 1 Aristotle (384 B.C.–322 B.C.)
The front cover image of the bust of Aristotle is used with express permission of the Acropolis Museum to whom our gratitude for this image is owed

In Memory of my devoted Mother Pili,
The advocate of my dreams

Someone who had begun to read geometry... asked Euclid, "But what advantage shall I get by learning these things?" Euclid called his slave and said, "give him three pence since he must needs make profit out what he learns."
– Euclid (Quoted by Stobaeus. *Extracts.* ii.31)

To seek from every kind of knowledge that some other thing come into being ... and that it must be useful is the act of someone completely ignorant of how great the distance is between things that are noble and those that are necessary.
– Aristotle. *Protrepticus (Frg. B42)*

Acknowledgments

At the outset I should wish to thank Dr. John Hymers and Ted Roedel for graciously reviewing my work and fruitful conversations on its topics and Dr. James Nicholson for his guidance on the texts of Sir Francis Bacon. In no way does their assistance signal agreement with my conclusions. I would like to thank my late father for nourishing my intellectual curiosity and fostering my education and my late brother for introducing me to philosophy. I would thank my many past professors who mentored me in my life's intellectual journey. I would mention Dr. Jean Loup Seban during my time at Princeton, and in Leuven Dr. Graham McAleer and Dr. Bart Raymaekers. These scholars by their guidance, and most of all by their example, helped me to discover and know the unique joy of a life to be found in a life devoted to the love of wisdom (φιλοσοφία). I would like to thank the editors at Springer in particular Neil Olivier and Diana Nijenhuijzen not only for their constant attentiveness in helping to bring this to practical fruition but also for taking on a project which is in one sense perhaps untimely (though I flatter myself to think in another sense it is *all too* timely). Thanks also are due to the reviewers for their helpful advice and suggestions. And finally, I would thank all my family and friends for their steadfast love and support, in particular my late mother to whom this present work is dedicated.

Contents

Part I The Emergence of the Theoretic Life in Classical Greek Philosophy

1 The New Barbarism: The Contest Between Classical Humanist Culture and Techno-Economic Pragmatism 3
References 9

2 In Pursuit of the Noble: The Classical Birth of the Liberal Arts 11
2.1 The Noble and the Useful Distinguished 12
2.2 The Hierarchy of Arts and Sciences 14
2.3 The Supreme and Most Noble of the Arts and Sciences 16
2.4 The Exaltation and Influence of the Liberal Arts 17
2.5 The Subordination of the Mechanical Arts 21
2.6 Humanism as the Center of the Classical Liberal Arts Tradition... 24
References 25

3 The Political and the Theoretic Life – The Challenge of Socrates 27
3.1 Nietzsche Against the "Theoretical Man" 27
3.2 The Political Nature of the Ethical Quest 29
3.3 The Good in Contention 31
3.4 The Rhetor-Statesman or the Philosopher-Statesman? 33
References 35

4 Aristotelian Teleology: The Bridge Between Natural Philosophy and the Problem of "The Good Life" 37
4.1 Aristotle and Plato 37
4.2 Natural Teleology 38
4.3 Man Within Nature – Is There a Natural Human Telos? 39
4.4 The Supreme Good 40
4.5 The Different Types of Life and the Question of the Best Life 41
4.6 The Life of Enjoyment (βίος ἀπολαυστικός) 43
4.7 The Function Argument 45
4.8 The Political Life (βίος πολιτικός) 46
References 47

xiii

5	**The Aristotelian Revolution: The Autonomy of the Theoretic Life and the Dream of Universal Science**	49
	5.1 The Theoretic Life (βιος θεωρητικός)	49
	5.2 The Substance of the Theoretical Life – The Theoretical Sciences	52
	5.3 The Role of Logic	52
	5.4 The Theoretical and Practical Sciences Distinguished	54
	5.5 The Division of Theoretical Science – Physics, Mathematics, and Theology	55
	5.6 The Universal Science	56
	References	57

Part II The Baconian Revolt Against Greek Theoria and the Modern Birth of the Technological Mind

6	**The Rebirth of Time: Sir. Francis Bacon and the Origins of Modernity**	61
	6.1 What Is Modernity?: The Birth of the Technological Mind	61
	6.2 Bacon's Place in History	64
	References	69
7	**Bacon's New Magic: The Transfigured Aim of the Sciences**	71
	References	75
8	**Technology Displaces Metaphysics – Bacon's New Hierarchy of the Arts and Sciences**	77
	References	80
9	**Breaking Aristotle's Bridge: The Modern Philosophical Critique of Teleology**	81
	9.1 Sir. Francis Bacon	81
	9.2 Rene Descartes	84
	9.3 Benedict Spinoza	87
	The Ethical-Political Implications of Modern Anti-Teleology	88
	References	92
10	**The Enlightenment as a Baconian Revolution**	93
	References	96
11	**Metaphysics Dethroned: Hume, Kant, and the "Self-Limitation of Reason"**	97
	11.1 David Hume's Skeptical Empiricism	99
	11.2 Kant's Critical Philosophy	102
	11.3 A Critique of the Critique: Kant's Anti-Metaphysical Revolution and Its Tension with Post-Newtonian Physics	105
	References	110

12	**Progressivism, Commerce, and the Triumph of Machine Civilization**	113
	12.1 The Idea of Progress	113
	12.2 The Valorization of Commerce and the Mechanical Arts	116
	12.3 Machine Civilization	118
	References	122
13	**The Classical Ideal of High Culture in the Democratic Age**	123
	References	128
14	**The Theoretic Life and the Challenge of American Pragmatism: Dewey and the Greeks in Contention**	129
	14.1 Progressivism	132
	14.2 Instrumentalism	133
	14.3 The Democratic Ideal: Is Liberal Education Anti-Egalitarian?	134
	References	137
15	**A Reply to Dewey**	139
	15.1 Progressivism	139
	15.2 Democratic Culture and Instrumentalism	141
	References	143
16	**The Contemporary Crisis of the Humanities: The Attack on the Western Canon and The Long Arm of Nietzsche, Marx, and Foucault**	145
	References	150
17	**The New Protrepticus: A Concluding Exhortation to the Theoretic Life**	153
	17.1 Philosophy and the Good Life –Happiness and Virtue	154
	17.2 The Tyranny of Utility: Pursuit of the Lower, Neglect of the Higher	156
	17.3 Reversing the Socratic Turn: The Restriction of Reason	158
	17.4 The Ideal of Universal Science vs. "The Barbarism of Specialization"	159
	References	162
Index		163

About the Author

Dr. Alexander S. Rosenthal-Pubul is currently an online Lecturer in political thought at Johns Hopkins University's Center for Advanced Governmental Studies from his home in Spain. He is also Co-Founder and Director of the Petrarch Institute which is dedicated to keeping alive the classical humanist tradition.

He holds a PhD in philosophy from Katholieke Universiteit Leuven in Belgium and an AB in philosophy from Princeton University in the USA. He has taught at a number of colleges and universities including the University of Glasgow, Loyola University (Maryland), and Catholic University of America (CUA). His main work has been at Johns Hopkins University where he has lectured since 2007 and served for a time as Assistant Director of the Center for Governmental Studies until 2012. Among his publications is his book *Crown Under Law* which deals with the intellectual origins of modern constitutionalism focusing on John Locke and Richard Hooker and their interaction with the medieval and renaissance traditions of political philosophy. His research interests include classical Greco-Roman, medieval and renaissance thought, European intellectual history, and political philosophy.

List of Figures[1]

Fig. 1	**Aristotle (384 B.C.–322 B.C.)** The front cover image of the bust of Aristotle is used with express permission of the Acropolis Museum to whom our gratitude for this image is owed	v
Fig. I.1	**Aristotle (384 B.C.–322 B.C.)** The front cover image of the bust of Aristotle is used with express permission of the Acropolis Museum to whom our gratitude for this image is owed. © Acropolis Museum NMA 5503, Photo: Socratis Mavrommatis ..	1
Fig. 3.1	*The Death of Socrates* by Jacques Louis David (1787) contributed by Everett -Art –www.shutterstock.com.........................	28
Fig. II.1	**Sir Francis Bacon (1561–1626)** Sir. Francis Bacon from a 1738 engraving – contributed by Everett Historical. – www.shutterstock.com	59

---

[1] **Note:** The first opening quote from Euclid is taken from Ivor Thomas's translation in the Loeb Classical Library Edition. *Greek Mathematical Works: From Thales to Euclid* (Harvard University Press, 1991 reprint):437 (citing Stobaeus. *Extracts* II.31.114 ed. Wachsmuth ii.228.25–29)

The second opening quote is from Aristotle's *Protrepticus*, B42 quoted in (A.W. Nightingale(ed.) (Cambridge University Press 2004) 194–195(Greek texts removed).

Part I
The Emergence of the Theoretic Life in Classical Greek Philosophy

Fig. I.1 Aristotle (384 B.C.–322 B.C.)
The front cover image of the bust of Aristotle is used with express permission of the Acropolis Museum to whom our gratitude for this image is owed. © Acropolis Museum NMA 5503, Photo: Socratis Mavrommatis

Chapter 1
The New Barbarism: The Contest Between Classical Humanist Culture and Techno-Economic Pragmatism

Modern culture stands under the menace of a new barbarism. "Barbarism" is the word which traditionally denotes those who are ignorant of, or who entirely reject, the edifice on which the whole Western tradition of high culture rests – the classical tradition of Greece and Rome. One might look to the Renaissance and think of Erasmus's *Libri Antibarbarorum*[1,2] which alludes to the enemies of Ciceronian humanism. By such a standard, our age, more even than that of Erasmus is becoming barbarized, ever more forgetful of its own intellectual and aesthetic inheritance.

The idea of a contemporary "crisis of the humanities" is not of course a novelty. Some decades back, Allan Bloom's *The Closing of the American Mind*[3] set off a national debate in America on the marginalization of the humanities within the university. At a time when major American schools like Stanford were removing their Western civilization requirements, Bloom's work formed part of a broader discussion

---

[1] Which can be found here: http://www.thelatinlibrary.com/erasmus/antibarb.shtml (July 1, 2017).

[2] The Renaissance humanists like Erasmus championed an education in polite or humane letters (*literae humaniores*) or good letters (*bonae literae*). This meant the study of the best Greek and Roman authors in order to initiate one into civilized habits of virtue, eloquence, civility and refinement. Conversely, ignorance or opposition to the classics was seen as the mark of "the barbarian". Particularly, it seems in Northern Europe, the rise of humanism met with much resistance – hence "the barbarians" addressed by the young Erasmus. The source of the term is the Greek βάρβαρος which in its most basic meaning simply meant a non-Greek. The Romans took over the term as the Latin *barbarus* which could refer anyone outside the orbit of Greco-Roman civilization, and naturally this had the additional connotation of being rude and uncultured. http://www.thelatinlibrary.com/erasmus/antibarb.shtml (accessed 9/28/2017). Insofar as this ideal of *humanitas* is closely associated with Cicero, it is essentially Ciceronian humanism which the Renaissance humanist movement was endeavoring to revive. See also James Tracy's article (1980) for background to this work of Erasmus. For another example of humanist counter-polemics against "barbarians", in this case those who resisted the introduction of Greek and humanist learning into England cf. also Roger Ascham's Letter to Erasmus: https://archive.org/stream/schor00asch/schor00asch_djvu.txt (Accessed May, 2018).

[3] Allan Bloom. *The Closing of the American Mind* (New York: Simon & Schuster, 1987).

on the place of the Western canon of Great Books in education. At the root of the objections to a Western core curriculum were certain theses, namely that culture and education are mere expressions of configurations of power, and hence that the Great Books approach is about cementing the position and privilege of a European civilization deemed to be malign and imperial. About this ideological dimension within the contemporary academy more will be said later.

Our contention however is that the crisis of the humanities has much deeper roots, reaching back to the origins of modernity itself. Let us consider basics. Our *zeitgeist* does not glory in philosophy, nor in arts and letters, but instead in material advancements, ingenious gadgetry, and economic acquisition. In general, the modern spirit is that of pragmatism and utility. Our modern form of education and of civilization as a whole, treats the human being as an instrument meant for other things. Technology in the service of economics – that is the sign of our age. Humanity is understood as *homo economicus* – "economic man".[4] We recognize and foster man[5] as technical inventor, man as consumer, man as investor, and man as an economically productive laborer, but *man as man*, is being laid to rest.

What is remarkable is how Western societies in late modernity have in this respect almost completely reversed the *Greek idea*. Edmund Husserl in his famous 1935 Vienna Lecture saw what he called "the theoretical attitude" as the distinguishing feature of Greek thought.

> Only with the Greeks, however, do we find a universal ("cosmological") vital interest in the essentially new form of a purely "theoretical" attitude.[6]

This "attitude" consisted precisely in a love for truth and wisdom as goods in themselves without any regard for practical concerns and consequences.

> The theoretical attitude, even though it too is a professional attitude, is thoroughly unpractical. Thus it is based on a deliberate *epoché* from all practical interests…[7]

This claim seems to be so far borne out by history. While for example the great civilizations of Egypt and Babylon developed ingenious mathematics for practical

---

[4] Economic materialism, a cause and effect of secularization as well, is as much a feature of Marxism which reductively identifies economic forces and relations as the motor of history, as of the prevalent attitude of "capitalistic consumerism" which sees in the accumulation of material wealth and comforts the primary or final end of life. In this respect, for all their mutual animus the latter day disciples of Adam Smith and Karl Marx often meet on common ground.

[5] It has become somewhat controversial in recent decades to use the term "man" or "mankind" to mean "the human being" on the premise that it is not gender neutral but privileges the male. The author is persuaded however this is an a-historical interpretation. In Old English "man" was completely gender neutral and one could attach the prefixes "wer" to designate a male or "wyf" to designate a female. The latter is the ancestor of "woman" while the former fell away. I am using the term in its original sense. No offence of course is intended. http://www.macmillandictionary-blog.com/his-and-hers-wyf-and-wer (accessed 8/30/17).

[6] From Husserl's Vienna Lecture –"Philosophy and the Crisis of European Man" translated by Quentin Lauer. http://people.ischool.berkeley.edu/~ryanshaw/nmwg/edmund.husserl-philosophy.and.the.crisis.of.european.man.pdf (Accessed August 28, 2017):164.

[7] Ibid. 168 (note – there is a different flat horizonal mark above the e in Epoche in the original.).

applications like architecture, land surveying, and trading, the Greeks like Pythagoras and Euclid developed rationalist mathematical systems for no reason beyond the contemplation of number and the intellectual beauty of deriving theorems and proofs from simple axioms. This idea of privileging the theoretic activities lies at the origin of the whole European tradition of the liberal arts which goes back to ancient Hellas. The liberal arts are those which have their own excellence and serve no practical end (as philosophy, poetry, music); and these were privileged over the "servile" or mechanical arts (as trade, agriculture, industry) which aim at mere utility.

Because the classical intellectual ideals have inspired supreme civilizational achievements, their loss or eclipse is a question of utmost moment for the future of our civilization. In tracing the intellectual archaeology of the revolt against the theoretical mind, we find that *the roots of the barbarism which threatens now to overtake us lie in the very origins of modernity itself.*

Why is this? We may here suggest only a provisional answer, an idea which will be further developed in the course of the present work. At the root of modernity lies a *change of aim*. This new aim differs from the old above all on the question of knowledge and its relation to life. For the Greek philosophers, wisdom is a good which is noble in itself; its possession is the highest perfection of the soul; its pursuit –philosophy – is the highest and most noble life for man.[8]

The first moderns reconceived knowledge as an *instrument*. Its aim is the augmentation of human power. Modernity then begins with a radical re-orientation – a revolution directed consciously against the classical conception. Whereas Aristotelianism had emphasized the dignity of knowledge and wisdom as *an end in itself*, - an end specific to man as the rational animal. The "moderns" radically reconfigured the aim of knowledge as *power*. The origins of this view can be localized in a single great historical personage – that of Sir Francis Bacon.[9] We will more fully deal with Bacon's revolution in thought in later chapters. Suffice it to say with his identification of knowledge with technical and useful power, *Baconianism is the first foundation of the whole modern project; it is the hinge on which all the rest rests.*

Modernity, that Baconian revolution, banished the old magic. But it envisioned science as a *new magic* which enables man to dominate natural forces and channel

---

[8] It is worth mentioning that Christianity introduced its own ideal of the highest life into Western culture – the Christian saint. Notable however is that the Greek conception was also largely incorporated (albeit in a less naturalist and more theological register) into the medieval Catholic world view. Hence St. Thomas Aquinas. (*Summa Contra Gentiles*.I.2) remarks that wisdom is the most perfect of human pursuits. But, Aquinas includes in concept not merely that knowledge and wisdom which is attainable through the exercise of natural human powers, but also at the highest level the knowledge of God made possible by grace which is consummated in heavenly life as the beatific vision of the divine essence. (ibid. 3.48-53) Hence in the Christian Aristotelianism of St. Thomas salvation is conceived largely as the supernatural perfection and fulfillment of the desire to know the highest Being.

[9] Francis Bacon. *The New Organon* Fulton H. Anderson (ed.).(New York: Macmillan Publishing Company, 1960):39(Aphorism III).

them to do his bidding. It cures illness, enables global communication and rapid transport, and creates wonders (though equally scientific technology confers power to destroy). In general, it aims at "the relief of man´s estate"[10] through facilitating concrete improvements to the human material condition. Modern civilization is a *technical civilization* oriented around *technical knowledge*.

One way then to approach this antithesis is in terms of the distinction - rooted already in the thought of the ancient Greeks - between technology (from τέχνη) and culture in the true and proper sense (παιδεία.) Modernity is characterized by the subordination of this classical ideal of culture to the technological ideal. The pragmatic, technical and utilitarian character of modern society today threatens the extinction of the classical ideal of the theoretic life. To examine what this ideal is, to examine why it is threatened, and to realize the value which is being lost – these things are our present subject.

The conflicting claims to primacy between culture and technology is the basis of the intellectual chasm dividing the classical from the modern. But what is the basis of the distinction itself between technology and culture? Technology is concerned with "the useful", much as culture is concerned with "the noble". One does not in the ordinary case use an iPad or telephone, or automobile, or some newly discovered miracle of medicine simply for the joy of it. They are *instruments* that help one to achieve some other goal as communication, transportation, or health– in other words they are *useful* to some further end beyond themselves.

Culture in this classical sense is a good of another order. One does not listen to a Haydn symphony, attend a Shakespeare drama, ponder Aristotle's *Metaphysics*, or gaze upon the Stanzas of Raphael for the sake of utility or profit. Indeed, such considerations might tend to corrupt the experience. These are things are meant to bring the human soul immediately to an experience of Truth and Beauty. The distinction between culture and technology is at its root a distinction between an *intrinsic* good and an *instrumental* one.

Each approach has given rise to distinct forms of education – there is *technical* education and *liberal* education.[11] A τέχνη is a practical craft or skill, so a technical education is an education designed to inculcate a set of particular skills needed to perform a particular work (as with those connected with gainful employment.) To learn how to farm, program a computer, or drive a truck would be all forms of technical education.

Education as *culture* by contrast aims at nothing beyond human excellence itself– the development of intellectual powers, moral virtues, the capacity to experience beauty. As H.I. Marrou put it:

> Classical teaching was chiefly interested in the man himself, not in equipping technicians for specialized jobs; and it is this, perhaps, that most sharply distinguishes it from the education of our own time, which makes its first aim to produce the specialists required by a

---

[10] Bacon. *The Advancement of Learning*. I.V.11.

[11] See for example H. Johnston "Liberal Education" in *The New Catholic Encyclopedia* (Washington D.C., CUA Press, 1967): Vol. 8, 700–701. Also see R.M. Ashley, "Liberal Arts", Ibid. 696–699.

civilization that to a fantastic extent has been invaded by technology and split up into fragments...it had no use for technique.[12]

The "technician" or "technical specialist" is indeed the characteristic human type of our time; much as "the gentleman" and "the universal man" were the characteristic types of Renaissance Europe, or the "monk" and "knight" were the characteristic types of the Middle Ages.

The ancients argued for the analogy of culture to agriculture.[13] Cicero tells us that:

> ...not all cultivated fields are productive...and in the same way not all educated minds bear fruit. Moreover, to continue the same comparison, just as a field, however good the ground, cannot be productive without cultivation, so the soul cannot be productive without teaching.[14]

This metaphor of agriculture is found also in the (pseudo) Plutarch. In the process of education, the mind of the student is the soil, the seed planted in the mind is the teaching, and the husbandman is the teacher.[15]

The point here is that true education aims at goods of the human soul, which like the soil must be *cultivated* if it is to bear fruit. Hence education is on this analogy the *cultura animi* – the culture of the soul. The conscious molding of man to an ideal of human excellence has been at the core of Western humanist education going back to the Sophists of ancient Greece. As Werner Jaeger avers:

> Protagoras claim that cultural education is the centre of all human life indicates that his education was frankly aimed at *humanism*. He implies that by subordinating what we now call civilization – namely technical efficacy – to culture : the clear and fundamental distinction between technical knowledge and power on the one hand, and true culture on the other, is the very basis of humanism.[16]

Cultura autem animi philosophia est.[17] – the culture of the soul is philosophy. From this audacious claim follows the conclusion that the philosophical life, or as Aristotle calls it the *theoretic life*, is the best form of life.

Now "philosophy" for the purpose of our discussion must be understood in the most capacious possible sense. As Cardinal Sadoleto, the great humanist educator noted:

---

[12] H.I.Marrou. *A History of Education in Antiquity*. ((Madison, WI: University of Wisconsin Press, from a 1956 copyright). Trans. George Lamb. Hereafter "Marrou".

[13] See the discussion in Werner Jaeger. *Paideia: The Ideal of Greek Culture, Vol. I* (Oxford University Press, 1962): 312–313. Hereafter Jaeger. *Paideia* I.

[14] *...ut agri non omnes frugiferi sunt...sic animi non omnes culti fructam ferunt. Atque, ut in eodem simili verser, et ager quamvis fertilis sine cultura fructuosus esse non potest, sic sine doctrina animus.*
Cicero. *Tusculan Disputations*. II. iv.13 in (Loeb Classical Library. London, William Heinmann, 1927):158–159 J.E. King Translator.

[15] Plutarch. *The Education of Children*. 4 http://penelope.uchicago.edu/Thayer/E/Roman/Texts/Plutarch/Moralia/De_liberis_educandis*.html (accessed 8/8/2016).

[16] *Paideia: The Ideal of Greek Culture, Vol. I* (Oxford University Press, 1962): 300.

[17] *Tusculan Disputations, supra* iv.,158.

> All knowledge or learning is liberal, but those arts of which we have so long been speaking, are, as it were, members of that one great body, philosophy itself...[18]

And so, the philosophical or theoretic life will embrace under its mantle all those arts and sciences which have their own proper excellence. It includes in the broadest sense every human study and activity which aims at knowledge of the Good, the True, or the Beautiful for no reason beyond the fact that these things are the highest and best.

The claim for the theoretic life has been approached from a number of angles. The *entire life of Socrates* may itself be considered an apology for the claim that "...the unexamined life is not worth living...".[19] Here, life and thought, the ethical and intellectual virtues, are conjoined as per the Socratic dictum that "virtue is wisdom."

Yet it was Aristotle who first established the *autonomy* of the theoretic life by distinguishing the intellectual from the moral virtues. If reason is both the distinguishing function of man and the highest and best thing in human nature, it follows that the life of reason will be the best form of human life. And the life of reason is expressed even more in philosophical contemplation than an active life devoted to political virtue.

In stark contrast to these thoughts, the "technological mind" is one which perceives of the value of any human activity with respect to profit and utility. From this standpoint the claims of the philosophers will appear to be a risible absurdity. From the standpoint of profit and utility it is difficult to argue that the efforts of metaphysicians to define the nature of Being is of greater utility or more profitable to man then are cooking, accounting, or computer science – or as one political figure recently put it "We need more welders, less philosophers."[20]

Such considerations are of course not new. For the practical man it has always been the *anti-utilitarianism* of philosophy that has appeared as its most salient feature. The antiquity of the objection indicates to what degree the type of "the

---

[18] Jacopo Sadoleto in *Sadoleto on Education. A Translation of the De Pueris Recte Instituendis* – Primary Source Edition. Ernest Trafford Campagnac (translator.) (Oxford University Press, 1916 – Nabu reprint): 126.

[19] Apology 38ª. Note on citations from Plato and Aristotle – Aristotle. *Metaphysics* (Loeb Classical Library, Harvard University Press, first print 1933). Hugh Tredennick (trans.) I am using Loeb Translations of Harvard University Press for most of the classical texts of Plato and Aristotle. In parenthesis will be the translator and the date of the first print. Those of Aristotle. includes the *Physics*.(Phillip K. Wicksteed 1929), the *Nicomachean Ethics*. (H. Rackham, 1999), *Politics* (H. Rackham trans, 1932) *and* of Plato the *Protagoras* (W.R.M. Lamb trans, 1924), *Meno* (same volume), *The Apology* (Harold North Fowler,1914), the *Phaedrus* (same volume.), the *Philebus* (Fowler 1925 first print.), and the *Gorgias* (W.R.M. Lamb first print 1925). Subsequent references will simply give the text, chapter and/or Bekker number for Aristotle or Stephanus number for Plato.

For *On the Soul*. I used Richard McKeon's translation in *The Basic Works of Aristotle* (New York: Random House, 1941).

[20] See commentary on Marco Rubio's statement here: http://www.weeklystandard.com/rubio-we-need-more-welders-and-less-philosophers/article/1062072 (Accessed September 22, 2017).

philosopher" has represented a perennial stumbling block for the type of "the common man" or "the practical man."

And yet the contemporary strain of pragmatism by making this perennial pragmatic disposition *normative* is from the perspective of history revolutionary. It completely overturns the traditional classical (and medieval) hierarchy of human activities. Where once the mechanical arts were subordinated, today it the liberal arts. What are the sources of the revolution which are bring down this classical ideal of *theoria,* and instead privileges the useful, pragmatic, and technical? What in short are the roots of modern anti-intellectualism?

This present work by seeking to answer this question is engaged in a capacious project of historical archaeology. Indeed, in searching for the roots of modern anti-intellectualism we will find in this work that they go deep into European history, into the roots of modernity itself with the Baconian rejection of the classical ideal. We will be providing a necessary exposition of the theoretic life itself. This Greek ideal regards a life devoted to the quest for wisdom as the most noble mode of human existence. It is for this inspiring and fertile ideal of human excellence that this work is an apology – if not perhaps an elegy.

References

Aquinas, St. Thomas. *Summa Contra Gentiles.* DHS Priory. https://dhspriory.org/thomas/ContraGentiles1.htm#2 and https://dhspriory.org/thomas/ContraGentiles3a.htm#50. Accessed May 2018.

Aristotle. 1941. *Basic Works of Aristotle.* Trans. Richard McKeon. New York: Random House.

Ascham, Roger. 1570. Letter to Erasmus in *Scholemaster.* John B. Major Ed. (1863) Archive.org. https://archive.org/stream/schor00asch/schor00asch_djvu.txt. Accessed May 2018.

Ashley, R.M. 1967. Liberal Arts. In *The New Catholic Encyclopedia,* vol. 8, 696–699. Washington, DC: CUA Press.

Bacon, Francis. 1605 (1893) *The Advancement of Learning.* Ed. Henry Morley. London: Cassell & Co at http://www.gutenberg.org/files/5500/5500-h/5500-h.htm. May 2018.

———. 1620 (1960). *The New Organon.* Fulton H. Anderson (ed.) New York: Macmillan Publishing Company

Bloom, Allan. 1987. *The Closing of the American Mind.* New York: Simon & Schuster.

Cicero. *Tusculan Disputations.* 1927 (translation) – London, Trans. William Heinmann. Loeb Classical Library. Harvard University Press.

Epstein, Ethan. 2015. Marco Rubio, Bad Guidance Counsellor. *The Weekly Standard.* http://www.weeklystandard.com/rubio-we-need-more-welders-and-less-philosophers/article/1062072. Accessed 22 Sept 2017.

Erasmus, Desiderius (c. 1480). *Libri Anti-Barbarorum.* The Latin Library http://www.thelatinlibrary.com/erasmus/antibarb.shtml. Accessed 28 Sept 2017.

Gough, Janet. 2011. His and Hers, *Wyf* and *Wer.* http://www.macmillandictionaryblog.com/his-and-hers-wyf-and-wer. Accessed 29 Aug 2018.

Husserl, Edmund. 1935. (1965 translation) "Philosophy and the Crisis of European Man" Trans. Quentin Lauer. New York: Harper http://people.ischool.berkeley.edu/~ryanshaw/nmwg/edmund.husserl-philosophy.and.the.crisis.of.european.man.pdf. Accessed 28 Aug 2017.

Jaeger, Werner. 1962a. *Aristotle: Fundamentals of the History of His Development.* Oxford: Oxford University Press.

———. 1962b. *Paideia: The Ideal of Greek Culture, Volume I*. New York: Oxford University Press.

Johnston, H. 1967. Liberal Education. *The New Catholic Encyclopedia. Vol 8*. Washington, DC: CUA Press

Marrou, H.I. *A History of Education in Antiquity*. Trans. 1982 (1956 copyright). Trans. George Lamb. Madison: University of Wisconsin Press.

Nightingale, A.W. 2004. *Spectacles of Truth in Classical Greek Philosophy*. Cambridge: Cambridge University Press.

Plato. 2001 (reprint). *Euthyphro. Apology. Crito.Phaedo. Phaedrus*. Ed. Jeffrey Henderson. Loeb Classical Library, Harvard University Press.

Plutarch. 1927. *The Education of Children*. F.C. Babbitt. Translator. Loeb Classics, Harvard University Press. http://penelope.uchicago.edu/Thayer/E/Roman/Texts/Plutarch/Moralia/De_liberis_educandis*.html. Accessed August 2016

Sadoleto, Jacopo. 1916 translation. (Nabu reprint.) *Sadoleto on Education. A Translation of the De Pueris Recte Instituendis* Trans. Ernest Trafford Campagnac (translator.) Oxford University Press, 1916 – Nabu reprint.

Tracy, James D. 1980. Against the 'Barbarians': The Young Erasmus and His Humanist Contemporaries. *The Sixteenth Century Journal* 11 (1): 3–22. https://doi.org/10.2307/2539472.

Chapter 2
In Pursuit of the Noble: The Classical Birth of the Liberal Arts

Image of Paris and Helen from the *Iliad* on a black antique plate contributed by Hoika Mikhail—www.shutterstuck.com

> *Quare liberalia studia dicta sint vides: quia homine libero digna sunt.*
> —Seneca. *Epistularum Moralem ad Lucillium (88)* http://www.thelatinlibrary.com/sen/seneca.ep11-13.3html (accessed May 5, 2018)

2 In Pursuit of the Noble: The Classical Birth of the Liberal Arts

> *Those studies are to be called liberal because they are worthy of free men.*
> –Seneca. *Moral Letter to Lucillius (88)*
>
> *Verily it is slavish to long for life, instead of for the good life… and to seek for money but pay no attention to the noble.*
> –Aristotle. *Protrepticus*. (From Aristotle's *Protrepticus* (fragment 52) cited in Werner Jaeger. *Aristotle: Fundamentals of the History of His Development*. (Oxford, 1962 -probable due to text wear): 60 (my ital.). Hereafter: Jaeger. *Aristotle*)

2.1 The Noble and the Useful Distinguished

There is today a standard and popular critique of philosophy, and more broadly of the liberal arts and liberal education. This critique may be summed up in the claim that they are "useless". From one point of view this is readily intelligible. The acquisition of technical *skills*, as for instance learning how to create and operate the computer programs which run factories and businesses. or for that matter how to use pliers or repair watches or fix indoor plumbing have obvious "marketability". These skills can be readily translated into jobs, profitability, and in general social utility. But what is the *use-value* of studying Virgil's poetry, the structure of a Bach fugue, or Aristotle's theories of speculative metaphysics?

This utilitarian critique rests of course on a hidden premise, namely that "utility" or "usefulness" is the proper standard of value. The perennial debate between "the philosopher" and "the practical man" stands or falls on an implicit claim of the latter, that utility is the highest form of human good, and consequently that utility or profit are the standard by which the value of philosophy (or any other human activity) are to be evaluated. Exploring how the ancient Greek philosophers met the popular objection of philosophy's "inutility" will take us to the very origins of the Western tradition of the liberal arts.

We need not be surprised about the antiquity of this kind of objection. Already in the *Clouds* of Aristophanes, the philosophers are lampooned as idle men who spend their lives fruitlessly contemplating idle things - such as the intestinal functions of gnats.[1] In fact, meeting this objection was the central concern of Aristotle's great (if today fragmentary) exoteric work the *Protrepticus*.[2] Hence it is with Aristotle that we begin.

---

[1] Aristophanes. *Clouds*. 153 http://www.perseus.tufts.edu/hopper/text?doc=Aristoph.%20Cl.%20153&lang=original (accessed April 25, 2018).

[2] Two excellent discussions which have informed my thinking on the Protrepticus and the general problem of *theoria* in Aristotle are in *Aristotle: Fundamentals of the History of His Development*. (Oxford, 1962). and more recently A.W. Nightingale(ed.) *Spectacles of Truth in Classical Greek Philosophy*, (Cambridge University Press, 20,045) Both have been invaluable to me in this work.

2.1 The Noble and the Useful Distinguished

At the heart of the Aristotle's examination of the question is evaluating what kind of good "the useful" is, and whether it can satisfy the claim that it is the only or the best kind of good. How does one go about evaluating which kinds of goods are better or worse?

A common procedure in Greek philosophy for determining the place of different goods in a hierarchy is to divide them according to where they are goods in themselves and or good for the sake of some other thing. For example in Plato's *Republic*, Glaucon asks Socrates if justice belongs to those goods which are desired for their own sake, or for the sake of the rewards one receives from being seen as just (or both.)[3] Likewise Aristotle in the *Nicomachean Ethics* in his search for the supreme good distinguishes between those ends or goods which are desired for their own sake of some further thing, and those which are desired for their own sake.[4] Those things which receive their goodness from the fact that they are necessary to attain some further good beyond themselves are ultimately only desirable for the sake of those things which are seen to have their own goodness.

On these grounds, Aristotle in the *Protrepticus* directly attacks the claim of utility to be the highest form of good:

> To seek from every kind of knowledge that some other thing come into being [from it] and that it must be useful is the act of someone completely ignorant of how great the distance is between things that are noble and those that are necessary; for the difference is vast. Those things that are loved for the sake of some other thing (which one cannot live without) should be called "necessities" and secondary causes, but those that are loved for themselves, even if no other thing results from them, should be called goods in the strict sense…It is completely ridiculous, then, to seek from everything a benefit beyond the thing itself and to ask "how is this profitable for us?" and "how is this useful"?'[5]

So, for Aristotle the claim that "the useful" is the highest form of good, flounders at once, as soon as we consider the very meaning of "utility." For something to be "useful" is precisely to be good for obtaining something beyond itself (e.g. e.g. exercise or medicine is "useful" for health). Merely useful things lack a proper excellence of their own but receive their value from the things they are useful for – as medicine receives its value from the higher good of health. Such goods are subordinate to higher ends, as medicine and exercise is a lower value than the health it is meant to produce (e.g. a medicine which actually harmed health would lack any value at all.) On this basis the Greeks were able to assert a hierarchy among goods.

While it seems at first paradoxical, it follows that we should expect then that the *highest* goods one can strive for will *not* be useful. These goods have their own

---

[3] Plato. *Republic*. Book II. 357Aff.

[4] Aristotle. *Nicomachean Ethics*. Book I.i (1094aff).

[5] From *Protrepticus*, B42 quoted A.W. Nightingale(ed.) *Spectacles of Truth in Classical Greek Philosophy,* (Cambridge University Press, 2004).194–195 (Greek texts removed – however the brackets are in the text. Save where mentioned as in this case, what is contained within brackets inside of quotations can be assumed to be added by the present author as for instance drawing attention to Greek terms employed in translated sections.]

value and so cannot be judged by their ability to help obtain anything beyond themselves. These goods belong to the *noble* (το καλόν) rather than to the merely necessary. This fundamental distinction between the noble, and the merely necessary or useful, helps to furnish the distinction between the liberal and mechanical arts, and to determine their relationship to each other.

2.2 The Hierarchy of Arts and Sciences

Each great art or science according to Aristotle aims at some end or good, as for instance medicine aims at health, or economics at wealth.[6] But by determining which of these are goods which are merely useful for things beyond themselves("necessary"), and which are goods in themselves ("noble" - i.e. good in themselves) it becomes possible to establish the *hierarchy of the arts and sciences*. As mentioned goods which serve to attain other goods are subordinate to the goods which they serve. Likewise, there are in many cases examples of sub-disciplines which are means to realize the end or good which is the province of another science or art (for instance the art of investing which involves managing equity portfolios is a means to the end of the science of economics which is wealth, and the art of building/construction serves the science of architecture). This utility of one science to another, is an indication for Aristotle that the science being served is a higher or master science(ἀρχιτεκτονῶν), and the one which serves its ends is the lower science for:

> …I say, that the ends of the master arts are things more to be desired then all those of the arts subordinate to them; since the latter ends are only pursued only for the sake of the former.[7]

In the opening of the *Metaphysics* Aristotle pursues a similar idea, but whereas the *Nicomachean Ethics* examines the hierarchy of the arts and sciences with respect to the idea of a practical science which is concerned with the goods which are to be pursued in actions and habits, the *Metaphysics* considers them in relation to wisdom of the understanding. Aristotle famously begins this work by establishing knowledge as a good desired for its sake.[8] There is however a hierarchy in forms of knowledge in that the more universal forms of knowledge (σοφία –or wisdom) which include knowledge of causes, are higher than practical experience (ἐμπειρία) of particulars which is based upon knowledge of particular effects – "For the experienced know the fact, but not the wherefore; but the artists know the wherefore and the cause."[9]

It is thus also possible to establish the hierarchy of the arts and sciences from this distinction between wisdom (knowledge of universal causes) and experience

---

[6] Nicomachean Ethics I.1ff
[7] Ibid. 4.
[8] Aristotle. *Metaphysics*. A 980ᵃ22.
[9] Ibid. 28–30.

2.2 The Hierarchy of Arts and Sciences

(knowledge of particular effects). The higher the science, the more it is universal, and the more it is concerned with knowledge of causes. This is true even where the person schooled in the higher art is less practically effective in the particular than the practitioner of a subordinate arts who knows their work from experience and habit– as for instance the architect may be less skilled in particular aspects of construction then the builder, the professor of medicine who may less efficacious in treating a particular patient then the nurse who attends that patient, and the economist may offer less prudent advice on which stocks to buy then the investor.

> For the same reason we consider that the master craftsmen in every profession are more estimable and know more and are wiser than the artisans, because they know the reasons of the things that are done...the master craftsmen are superior in wisdom, not because they can do things, but because they possess a theory and know the causes.[10]

A final criteria Aristotle provides for discerning the hierarchy of the arts and sciences is that of leisure as opposed to labor. The topic of leisure which the Greeks called σχολή and the Romans *otium* was itself a topic of profound importance to classical life and reflection. Central to the classical view then is an evaluation of *labor* (whether manual or commercial) and its distinctively subordinate status to *leisure*. By this is certainly not meant what we might think of today - slothfulness or mere indulgence in pleasurable entertainments. The Greeks and Romans took leisure *seriously*. We may think of Cicero's quest for consummating his life of public duty with periods of *otium cum dignitate* (honorable leisure) to which he could devote himself to philosophy; of Seneca's treatise *De Otio* (on leisure); or of Augustine's retreat to Cassicacum for an extended period of reflection with friends upon his Christian conversion. The problem of leisure and labor is implicated in the whole debate between the active and the contemplative life which was taken up by such luminaries as St. Thomas Aquinas and Petrarch.

Aristotle's argument for the superiority of leisure to labor should already be familiar. Labor is merely useful, while leisure is a good in itself. Indeed, the labor is related to leisure as means to end, for one engages in labor in order to have leisure.[11] Labor in short is concerned with "the useful" as leisure in the highest sense is focused on "the noble."

The higher arts and sciences correspondingly will be those which exist and were invented for their own sake then those which have merely useful aims and were invented for the purposes of supplying man with necessities. One might think of agriculture, trade, building, tailoring, etc.... few of which would be done for their own sake if they did not supply human needs for food, clothing, and shelter. Aristotle speculating on the origins of arts and sciences notes and its connection with leisure (and leisured classes) he notes:

> It is therefore probable that at first the inventor of any art which went further than the ordinary sensations were admired by his fellow-men, not merely because some of his inventions were useful, but as being a wise and superior person. And as more and more arts were dis-

---

[10] Aristotle. *Metaphysics*. A 981a-b.
[11] Aristotle. *Nicomachean Ethics*. X.vii (1177b).

covered, some relating to the necessities, other to the pastimes of life, the inventors of the latter were always considered wiser than those of the former, because their branches of knowledge did not aim at utility...the mathematical sciences originated in the neighborhood of Egypt because there the priestly class was allowed leisure.[12]

From this exposition, we have seen that Aristotle provides three criteria for determining the place of the various arts and sciences within the hierarchy. First, those sciences which concern themselves more with the universal cause are higher than those which concern themselves with particular causes or with effects. Secondly, the higher sciences are more "leisured" because they have their own proper excellence, as opposed to those which serve a good other than themselves (such as supplying the necessities of life). Finally, those sciences whose good or end lies in their utility for other sciences are subordinate relative to those sciences. In short, those sciences whose ends are noble and good in themselves are superior to those whose ends are merely useful or are good for the sake of other things. Let us examine this in more detail.

2.3 The Supreme and Most Noble of the Arts and Sciences

But if there is a hierarchy of the arts and sciences, is there also a *supreme* science which sits atop the hierarchy? Aristotle will argue that such an art is philosophy, more particularly in its theoretical and foundational role as *first philosophy* (what was later called metaphysics.) But what are his grounds for this claim? And what are the attributes of this science?

From the foregoing considerations, we have seen that it is possible for Aristotle to adduce the characteristics of this science. The first as we have touched on is that this will be a science which concerns itself with the most universal forms of knowledge and the most universal causes.

> For the man who desires knowledge for its own sake will most desire the most perfect knowledge, and this is the knowledge of the most knowable, and the things which are most knowable and the things which are most knowable are first causes and principles...[13]

Secondly, is the criterion of leisure. This knowledge sought by this science must be motivated purely by leisured curiosity, and not by any concern for utility and necessity. Aristotle will characterize the motivation of philosophy in terms of *wonder* (θαυμάζειν).

> That this is not a productive science is clear from the considerations of the first philosophers. It is through wonder that men now begin and originally began to philosophize...if it is to escape ignorance that men studied philosophy, it is obvious that they pursued science for the sake of knowledge, and not for any practical utility. The actual course of events bears

---

[12] Aristotle. *Metaphysics*. A 981b.
[13] Ibid. 982b.

witness to this; for speculation of this kind began with a view to recreation and pastime, at a time when practically all the necessities of life were already supplied.[14]

Thirdly, and closely related to this, this science must serve no end beyond itself. Every art and science which is useful for another science, for Aristotle is in a sense a servant, since its value comes from the service it provides to another branch of knowledge.

The most noble of the arts and science, the "master science" having the supreme excellence will serve no other, but rather will be served by the other arts and sciences which are subordinate to it. It is the "master" science and the others are merely its servants which exist for its sake and not their own:

Again among the sciences we consider that science which is desirable in itself and for the sake of knowledge more nearly Wisdom that that which is desirable for its results, and that the superior is more nearly Wisdom than the subsidiary...[15]

Such a science will be like a free man who exists for himself and serves no one else.

...for just as we call a man independent who exists for himself and not for another, so we call this the only independent science since it alone exists for itself.[16]

A consequence of this consideration, ironic and so perplexing to the modern mentality, is that so far from detracting from its value, *the apparent inutility of philosophy is actually evidence of its supreme nobility:*

Accordingly, although all other sciences are more necessary than this, none is more excellent.[17]

Aristotle central metaphor is that the supreme science -First Philosophy – then is as that of the free man who is the master of the other sciences. In comparison with the "freedom" and "independence" of this master science who serves nothing beyond itself, the other sciences are "servile" aiming not at their own good, but as the good of the master. Aristotle's metaphor provides the foundation for the broader classical idea of the liberal or "free" arts which find their value not principally in their utility, but in their intrinsic nobility and excellence.[18]

2.4 The Exaltation and Influence of the Liberal Arts

The educational tradition of the liberal arts based around the Greek and Roman classics has played a singular role in the history of Western civilization. Like the Confucian tradition of China which was also based around a set of classical texts, it

---

[14] Idem.
[15] Ibid. 982a.
[16] Aristotle. *Metaphysics*. A 982b.
[17] Ibid. 983a.
[18] See H. Johnston. "Liberal Education". *New Catholic Encyclopedia*. Volume 8. (Washington D.C.: CUA Press, 1967, 1981 reprint):700–701.

long represented social capital, knowledge of which was an asset to advancement within European societies, as well as an educational tradition whose transmission preserved this civilization's identity across time. As the British historian Christopher Dawson put the matter:

> For the system of classical studies or "humane letters" …had its origins some twenty-four centuries ago in ancient Athens and was handed down intact from the Greek sophists to the Latin rhetoricians and grammarians, and from there to the monks and clerks of the Middle Ages. These in turn handed it on to the humanists and school-masters of the Renaissance from whom it finally passed to the schools and universities of modern Europe and America.[19]

What is a liberal art? Although the precise formalization of the liberal arts (*artes liberales*) seems to occur under the Romans, the basis of the idea can be found in what has already been discussed in Aristotle. Just as the "free" gentleman does not serve another, so the highest arts will be "free" in the sense of desirable for their own sake. Hence the origin of the distinction between "servile arts" which are "merely" useful, and the liberal arts which are excellent in and of themselves. Seneca provides us with one of the clearest explanations of the meaning of the "liberal arts" when he writes in his letter to Lucillius that *Quare liberalia studia dicta sint vides: quia homine libero digna sunt.*[20] "Those studies are to be called liberal because they are worthy of free men" The "free" man here is a reference to the gentleman who is free from the burden of the merely necessary. Hence, an aristocratic ideal enters into the very conception of the liberal arts. The person who is freed from the burden of needing to labor for their necessities owns their own time – hence they are a "free man". Hence such a man is "free" to pursue those arts which are noble, dignified, and truly worthy of man. However, we must not think that the ancients were thinking simply in crassly economic class terms. Epictetus had been a slave, and Socrates, poor and from a background of manual labor, is praised by Xenophon as the great exemplar and teacher of gentlemanly virtue.[21] And as we shall see the excessive concern for and pursuit of money was generally viewed as vulgarizing.

Which arts can be considered "liberal"? There has always been some debate. In virtually all iterations philosophy (which was considered by Aristotle to be the only art which is not servile) belongs to the liberal arts. Yet the conception of the liberal arts has always been substantially broader than that. Poetry has generally been included, perhaps because of its leading role in the so called old Greek system of education (ἡ ἀρχαία παιδεία).[22] The two pillars of this system were gymnastic to form and perfect the body, and poetry to form and perfect the soul. As Jaeger notes,[23] for the Greeks, poetry is valued not only for the profound aesthetic experience it can

---

[19] Christopher Dawson. *The Crisis of Western Education.*(Washington D.C.: CUA press, 2010): 5.
[20] Idem.
[21] See for example Xenophon's *Memorabilia*. 2.2-3.
[22] This term seems to come from Aristophanes. See H.I. Marrou (supra), Chapter IV, 36ff. for the reference and general discussion of the archaic Greek education.
[23] See for example Jaeger. *Paideia* I, Chapter III "Homer as Educator".

2.4 The Exaltation and Influence of the Liberal Arts

produce, but for its specifically *educational* function, meaning here the education of the soul. The oldest and greatest of the Greek poets, Homer is called by Plato τὴν Ἑλλαδα πεπαίδευκεν[24] - "the educator of Greece." From the example of Homeric heroes, the Greeks first learned the value of competitive striving for ἀρετή[25] (excellence or virtue) and the prize of honor which it justly claims. This is evinced in the constant references to Homer (among many other poets) in their works. Hence for example Socrates in his *Apology* refers to the example of Achilles on why it is base to avoid death at the cost of dishonor.[26]

If it was from poetry that the Greeks learned about an active life of honor expressed in noble deeds, the pursuit of such a life in the deliberative assemblies of the polis required new arts. Hence, beginning with the sophists, rhetoric also played a significant role in the liberal arts which indicates also the fact that an art being "worthy of a free man" did not altogether exclude practical concerns. It was the responsibility of free citizens to participate in public life. With the Romans in particular, the active and public life of duty and honor was also seen as especially worthy. This kind of honorable "political life" one sees exemplified in the life of Cicero.[27] And so, in the context of societies like Greece and Rome where political decisions are made by assemblies and senates, the use of words and the power of persuasion were essential. Since the Sophists and Isocrates at least and culminating in works like the *Institutio Oratoria* of Quintilian, rhetoric became in fact the foundation of a system of liberal arts education for the rhetor-statesman must be broadly educated in order to speak on any theme.

Nor were the natural sciences and mathematics neglected. An influential formalization of the liberal arts occurred in late antiquity in the early fifth century with Martianus Capella's work *De Nuptiis Philologiae et Mercurii* on which the medieval based their trivium focused on language – (Latin) grammar, rhetoric, and logic and the mathematical-scientific quadrivium of arithmetic, geometry, music, and astronomy. Capella's system of the seven liberal arts based around the study classical authors and texts formed the foundation of medieval monastic and university education. Typically for medieval man the role of the liberal arts was preparatory – one passed through the training in the faculty of arts before moving on to the professional higher faculties of medicine, law, and theology.

With the Italian humanists and teachers of the Renaissance court schools in the fifteenth century a number of shifts took place. The restoration of Greek to the West thanks to Byzantine scholars like Bessarion, Chrysoloras, and Gemistus Pleithon who moved to Italy, as well as eager students like Marsilius Ficino of the Florentine academy had a profound impact on intellectual life. The study of antiquity was also

---

[24] Plato. *Republic* X. 606.

[25] Cf. for example Homer. *Iliad*. VI. 206.

[26] Plato. *Apology*. 28b-d.

[27] This Ciceronian ideal of the active political life was taken up in the "civic humanism" of the Florentine Renaissance.

no longer seen as principally preparatory to theology, but as part of broad humanist development. As Christopher Dawson puts the matter:

> ...the humanists learnt from Xenophon, and Plato, and Isocrates that education is an art which should aim at the harmonious development of every side of human nature, physical, moral, and intellectual. Thus they gained an awareness of the unlimited possibilities of the enrichment of personal life by art and literature and social intercourse.[28]

Moreover, education rather than being largely professional training for clerics, became more a matter for the laity. Hence if one compares the liberal arts in the form of the Renaissance *studia humanitatis* with the trivium and quadrivium of the Middle Ages one notices both overlaps and differences. The *studia humanitatis* associated with figures like Leonardo Bruni typically included grammar, rhetoric, poetry, history, and moral philosophy.[29] The logic so prized by the medieval was de-emphasized in favor of rhetoric regained in the context of the Italian city states like the Florentine republic. History was added to standard study of "humane letters", as was poetry which reflected the strong literary and aesthetic concerns of the renaissance. The renaissance court schools which were centers of humanism transmitted really Seneca's concept of the liberal arts as those suitable to the free gentleman which we have seen already. Pier Paolo Vergerio one of the most important figures in this regard expressed the ideal succinctly:

> We call those studies liberal, then, which are worthy of a free[*liber*] man: they are those through which virtue and wisdom are either practiced or sought, and by which the body or mind is disposed toward all the best things...Just as profit and pleasure are laid down as ends for illiberal intellects, so virtue and glory are goals for the noble.[30]

The *Roman* aspect of the Italian renaissance culture is seen in the emphasis on the liberal arts as suitable to the active life of public duty rather than merely to philosophical (or monastic) contemplation. The Renaissance conceptions were however by no means merely imitative. In that regard a particular credit of Renaissance, perhaps especially the fifteenth century Italian Renaissance figure Leon Battista Alberti to essentially elevate the three fine arts of painting, sculpture, and architecture to the rank of liberal arts, which he did in his trinity of works devoted to each - *De Pictura, De Statua,* and *De Re Aedificatoria.*[31] One notices that for the Greeks none of these arts are represented among the Muses,[32] perhaps reflecting the fact that while music and poetry in their different forms (epic, lyric, and dramatic) were

---

[28] Dawson. *The Crisis of Western Education.* (Supra), 24. Chapter III contains an excellent summary of education in the Renaissance.

[29] Leonardo Bruni. *The Study of Literature.* Trans. Craig W. Kallendorf. In *Humanist Educational Treatises.* (Cambridge, Massachusetts: Harvard University Press, 2008).

[30] Pier Paolo Vergerio. *Character and Studies.* Ibid. 12. (Brackets with Latin term in original)

[31] See Leon Battista Alberti "On Painting". http://www.noteaccess.com/Texts/Alberti/index.htm and *De Re Aedifatoria.* 1755. London: Edward Owen, http://archimedes.mpiwg-berlin.mpg.de/docuserver/images/archimedes/alber_archi_003_en_1785/downloads/alber_archi_003_en_1785.text.pdf (Accessed May, 2018).

[32] I believe this fact is discussed in this BBC talk https://www.bbc.co.uk/programmes/b07bft7v (accessed April 28, 2018).

given high status, painting, sculpture, and architecture were deemed perhaps "tainted" by their association with manual labor and artisanship. But the consequent glorious flowering of renaissance art witnessed in such works the sculpture of Donatello and Ghiberti, the painting of Botticelli, Raphael, and Da Vinci, the architecture of Brunelleschi and Bramante, and Michelangelo's supreme achievements in all three arts bear witness to the myopia of this view. For while in the form of work done arts like sculpture and painting share traits in common with "mechanical" arts, their aim is not primarily utility but aesthetic contemplation. In seems therefore that the most salient distinction between liberal and mechanical arts is not to be found in the genus of activity, but rather in its intentions of the activity, as to whether their ends are noble or merely useful. What remains unsettled is whether the aim of the liberal arts is to fit someone for an active life of public virtue or a theoretic life of leisured contemplation. For Cicero it seems the answer would be both - their supreme value lies in their benefits in all moments and walks of life. In his *Pro Archia* he writes of liberal study that:

> …even if there were no such great advantage to be reaped from it, and if it were only pleasure that is sought from these studies, still I imagine you would consider it a most reasonable and liberal employment of the mind: for other occupations are not suited to every time, nor to every age or place; but these studies are the food of youth, the delight of old age; the ornament of prosperity, the refuge and comfort of adversity; a delight at home, and no hindrance abroad; they are companions by night, and in travel, and in the country.[33]

2.5 The Subordination of the Mechanical Arts

We have seen the classical exaltation of the liberal arts as those having their own proper excellence and hence worthy of the "free gentleman." Such is one who being free from necessity, owns his own time and hence can pursue noble things. Corresponding to this conception would be the subordination of the merely necessary arts as lower in the hierarchy of human activities. These illiberal or "servile" arts would be those which have no good in themselves but are merely useful to other ends. Clearly such arts would be subordinate in value to the liberal arts.

In fact, it seems to be only in the Middle Ages that efforts were made to categorize and enumerate them with any precision. The medieval scholars commonly contrasted the mechanical arts (*artes mechanicae*) with the liberal arts (*artes liberales*) which made up the foundation of the medieval educational tradition. As the liberal arts were organized into 7 – the trivium (grammar, logic, and rhetoric) and qua-

---

[33] Cicero. *Pro Archia Poeta*. Quod si non his tantus fructus ostenderetur, et si ex his studiis delectatio sola peteretur, tamen (ut opinor) hanc animi adversionem humanissimam ac liberalissimam iudicaretis. Nam ceterae neque temporum sunt neque aetatum omnium neque locorum: haec studia adulescentiam alunt, senectutem oblectant, secundas res ornant, adversis perfugium ac solacium praebent, delectant domi, non impediunt foris, pernoctant nobiscum, peregrinantur, rusticantur. (http://www.forumromanum.org/literature/cicero/arche.html#16 – accessed April 29, 2018). http://www.forumromanum.org/literature/cicero/arch.html#16

drivium (arithmetic, geometry, music, astronomy), so they often tended to catalogue the mechanical arts in a corresponding way. One of the most interesting medieval elaborations of the mechanical arts is found in Hugh of St. Victor's eleventh century work the *Didascalicae* which provides seven mechanical arts as analogues to the seven liberal arts – *lanificium* (weaving), *armaturum*(armoring), navigationem (navigation), *agriculturam* (agriculture), *venationem* (hunting), *medicinam* (medicine) and *theatricum* (the theatre).[34] All these manifold arts are distinguished from liberal arts by the fact that they aim *de opere artificis agunt quod a naturam formam mututatur*[35] that is to say by the operation of the worker they change natural forms, and *fabricae omnium*[36] - make all things. In short, the mechanical arts unlike the liberal arts are productive aiming at the making of things. The mechanical arts thus have a connection with our modern ideas of technology and technical utility, since these activities like medicine, manufacturing, medicine, and making weapons are among the most areas of development in modern technology.

In spite of the late categorization, the classical ancestors of the medieval idea of "mechanical arts" are clear enough. The Greeks designated illiberal, merely useful arts by the somewhat derogatory term βάναυσος – "mechanical." Such arts were generally held to be vulgar and un-gentlemanly by both the philosophers and the Greek aristocracy. As Xenophon puts the matter

> The illiberal [βαναυσικαί] arts, as they are called, are spoken against, and are, naturally enough, held in utter disdain in our state."[37]

The reasons given are that the excessive occupation with mechanical labor leaves little time for the cultivation of the mind, or for generous service to one's friends and city.

While it is doubtless true there is a strong "aristocratic" element in these evaluations we should be careful to associate these distinctions simply with possession of money and property or its absence. Xenophon for instance characterizes Socrates, famous for his poverty, as the very model of a "gentleman" (καλὸν κἀγαθὸν).[38] It should also be emphasized that the disdain for the mechanical arts extended not merely to manual labor but also to commerce especially by those who occupy their minds with the vulgar pursuit of wealth, and so are taken away from intellectual and moral culture. Take for instance Plato's portrait of the "oligarchic

---

[34] Hugh of St. Victor. *Erudutionis Didascalicae. Libri Septem. (Liber Secundus. XXI)*, P. 760 http://www.documentacatholicaomnia.eu/02m/1096-1141,_Hugo_De_S_Victore,_Eruditionis_Didascalicae_Libri_Septem,_MLT.pdf – (from Migne Patrologia - accessed 5/4/2017).

[35] Idem.

[36] Idem.

[37] Xenophon. *Oecenomicus*. IV. 2. http://www.perseus.tufts.edu/hopper/text?doc=Perseus%3Atext%3A1999.01.0212%3Atext%3DEc.%3Achapter%3D4%3Asection%3D2 (accessed 3/25/2015). Greek in brackets added.

[38] Xenophon. *Memorabilia*. I.2.17 http://www.perseus.tufts.edu/hopper/text?doc=Xen.+Mem.+1.2.17&fromdoc=Perseus%3Atext%3A1999.01.0207 (Accessed May 5, 2018). The term implies a combination of noble beauty (as of the soul), as well as moral goodness.

2.5 The Subordination of the Mechanical Arts

man" who rules the "oligarchic polis.". In a dialogue between Socrates and Glaucon, Socrates begins:

> "He would be a squalid fellow" said I, "looking for a surplus of profit in everything, and a hoarder, the type the multitude approves. Would this not be the character of the man who corresponds to such a polity?"" I certainly think so", he said. "Property at any rate, is the thing most esteemed by that state and that kind of man." "That I take it" said I, "is because he has never turned his thoughts to true culture[παιδείᾳ]…Shall we not say that owing to this lack of culture [ἀπαιδευσίαν][39] "the appetites of the drone spring up in him…?"

The deprecation of both manual labor and commercial profiteering was easily inherited by Roman gentlemen. Hence Cicero tells us that:

> Unbecoming to a gentleman, too, and vulgar are the means of livelihood of all hired workmen whom we pay for mere manual labour, not for artistic skill; for their case the very wage they receive is a pledge of their slavery. Vulgar we must consider those who buy from wholesale merchants to retail immediately…And all mechanics are engaged in vulgar trades; for no workshop can have anything liberal about it.[40]

Or Consider the view of Seneca:

> …I respect no study, and deem no study good, which results in money-making. Such studies are profit-bringing occupations, useful only in so far as they give the mind a preparation and do not engage it permanently. One should linger upon them only so long as the mind can occupy itself with nothing greater.[41]

All the foregoing re-enforces the basic point that classical world view implied a definite hierarchy in which the liberal arts concerned with intellectual, moral, and aesthetic cultivation are superior to the mechanical arts -those concerned with mere utility or the acquisition of wealth. As man is higher than material goods, and as the soul is superior to the body, so are the arts that pertain that to the excellence of the soul are innately higher than those concerned with the excellence of the body and certainly that those concerned with the acquisition of material goods.

---

[39] Plato. *Republic*. 554a-554c. (my brackets).

[40] "Illiberales autem et sordidi quaestus mercennariorum omnium, quorumoperae, non quorum artes emuntur; est enim in illis ipsa merces auctoramentum servitutis. Sordidi etiam putandi, qui mercantur a mercatoribus, quod statim vendant; …Opificesqueomnes in sordida arte versantur; nec enim quicquam ingenuum habere potest officina. Minimeque artes eae probandae, quae ministrae sunt voluptatum":Cicero. *De Officiis* I. 151. http://www.perseus.tufts.edu/hopper/text?doc=Perseus%3Atext%3A2007.01.0048%3Abook%3D1%3Asection%3D150 (Accessed 1/3/2018).

[41] Lucius Annaeus Seneca, *Epistulae Morales*, LXXXVIII nullum suspicio, nullum in bonis numero quod ad aes exit. Meritoria artificia sunt, hactenus utilia si praeparant ingenium, non detinent. Tamdiu enim istis inmorandum est. quamdiu nihil animus agere maius potest; (http://www.thelatinlibrary.com/sen/seneca.ep11 13.shtml - accessed April 29, 2018) English translation; *Moral Letters to Lucillius: Letters from a Stoic. Aegitus.* Can be accessed here: https://play.google.com/store/books/details/Seneca_Lucius_Annaeus_Moral_letters_to_Lucilius?id=6ykJAwAAQBAJ. (*Accessed October 25,2018*). *I believe this is the Richard Mott Gumere translation.*

2.6 Humanism as the Center of the Classical Liberal Arts Tradition

The classical subordination of mechanical arts involved in manual labor and commerce to the liberal arts, runs up against all the tendencies of modern culture from its economic, pragmatist and technical orientation to its democratic-egalitarian sympathies. But whatever may be legitimately conceded to the modern, Marxist tinged critique of the liberal arts tradition for its class roots in ancient slave societies, we are apt by focusing on these contingent sociological facts to miss the basic *humanism* that underlies the Greco-Roman world view as expressed by its most formidable intellectual representatives. What is humanism in the classical sense? Its origins and fundamentals are found in this aforementioned Greek ideal of παιδεία. We can understand it as "education" provided we understand its aims correctly. It should already abundantly clear that we are not thinking of "education" in the sense "training" to learn the techniques and skills proper to a profession. Rather its goal is as Jaeger wrote "…the process of educating man into his true form, the real and genuine human nature[42]" or as Marrou put it to achieve "the mind of a man who has become truly man."[43]

To use a somewhat Aristotelian analogy, just one imagines that the seed has within it the potency to fully actualize its nature by blossoming into a majestic, full grown flower, so human beings have the potency to become fully human – i.e. perfecting the full range of human powers. Save that this process of becoming fully *human* is not an automatic organic process and but requires education and culture – παιδεία -which the Romans (having no comparable word in the Latin tongue) fittingly translated as *humanitas* to express the educational goal of a fully developed humanity.

One of the basic roots of the concept is found in Aulus Gellius's *Attic Nights*

> Those who have spoken Latin and have used the language correctly do not give to the word humanitas the meaning which it is commonly thought to have, namely, what the Greeks call φιλανθρωπία, signifying a kind of friendly spirit and good -feeling toward all men without distinction, but they gave to humanitas about the force of the Greek παιδεία; that is what we call eruditionem institutionemque in bonas artes, or "education and training in the liberal arts." Those who earnestly desire and seek after these are most highly humanized. For that pursuit of knowledge, and the training given by it, and been granted to man alone of all the animals, and for that reason it is termed humanitas…[44]

The ideal of the liberal arts then is that they aim at cultivating what is *particular* to man and distinguishes him from the animals - his intellectual and moral faculties. They differ from the mechanical arts which at the acquisition of goods, while the liberal arts aim at the *cultivation of man himself,* in other words the development of

---

[42] Jaeger. Paideia I. xxiiii.
[43] Marrou.99.
[44] Aulus Gellius. 13:17 http://penelope.uchicago.edu/Thayer/E/Roman/Texts/Gellius/13*.html#17 (May 5).

the full range of human powers. The superiority of the liberal arts therefore is rooted in nothing beyond the superiority of man as an intellectual being to material goods.

For Aristotle, it is important to recall that human nature itself has a "slavish" element insofar as the body has necessities and everyone is compelled to look to its upkeep. Hence, he remarks that "…in many respects human nature is servile…".[45] But as man is a greater thing then his goods, so the education which perfects man himself is higher than that which teaches him how to acquire such goods. The excellence of the soul, that is virtue and wisdom, are much worthier of aspiration then the material goods that may adorn us. As he puts it:

> Believe that man's happiness lies not in the magnitude of his possessions but in the proper condition of his soul…only the cultivated soul is to be called happy; and only the man who is such, not the man who is magnificently decorated with external goods, but is himself of no value.[46]

It is from such an ideal of *humanitas* that we derive the *studia humanitatis* of the Renaissance, and the "humanities" and "humanism" of later times. Through the study of the best authors of Greek and Roman antiquity, they aimed specially to achieve the excellence of the soul – virtue and wisdom. Renaissance educators expressed the aim of humane letters in this way which has become an imperishable ideal:

> We receive from Nature what is central in ourselves, what makes us truly and individually what we are, but in a rough and unfinished form; it is the function of letters to bring this to its highest perfection and to work out in it a beauty comparable to its divine original.[47]

References

Alberti, Leon Battista. *On Painting (De Pictura)*. At Noteaccess.com. http://www.noteaccess.com/Texts/Alberti/index.htm. Accessed May 2018.
———. *De Re Aedifatoria*. 1755. London: Edward Owen. http://archimedes.mpiwg-berlin.mpg.de/docuserver/images/archimedes/alber_archi_003_en_1785/downloads/alber_archi_003_en_1785.text.pdf.
Aristotle. Nicomachean Ethics. 1926. (1999 reprint). *Nicomachean Ethics*. Trans. H. Rackham, 1999. Loeb Classical Library, Harvard University Press.
———. *Metaphysics* 1933 (2003 reprint). Trans. Hugh Tredennick. Loeb Classical Library, Harvard University Press.
Aristophanes. *Clouds*. In Perseus. http://www.perseus.tufts.edu/hopper/text?doc=Aristoph.%20Cl.%20153&lang=original. Accessed 25 Apr 2018.
BBC. 2016. *The Muses*. In Our Time. https://www.bbc.co.uk/programmes/b07bft7v. Accessed May 2018.

---

[45] Aristotle. Metaphysics. A 982b.
[46] Proptrepticus. Frg. 51 Cf Jaeger *Aristotle, 51*.
[47] Jacopo Sadoleto in *Sadoleto on Education. A Translation of the De Pueris Recte Instituendis – Primary Source Edition*. Ernest Trafford Campagnac (translator.) (Oxford University Press, 1916 – Nabu reprint): 12.

Capella, Martianus. 1836. *De Nuptiis Philologiae, et Mercurii*. Francofurti ad Moenum Varentrapp. At Archive.org. https://archive.org/details/denuptiisphilolo00martuoft. Accessed May 2018.

Cicero. Pro *Archia Poeta*. In Forum Romanum. http://www.forumromanum.org/literature/cicero/arche.html#16. Accessed 29 Apr 2018.

———. Cicero. *De Officiis*. In Perseus. http://www.perseus.tufts.edu/hopper/text?doc=Perseus%3Atext%3A2007.01.0048%3Abook%3D1%3Asection%3D150. Accessed 3 Jan 2018.

Dawson, Christopher. 2010 (reprint). *The Crisis of Western Education*. Washington, DC: CUA Press.

Gellius, Aulus. *Noctes Atticae*. Transcribed from Loeb 1927 (revised 1946). http://penelope.uchicago.edu/Thayer/E/Roman/Texts/Gellius/1*.html. Accessed 5 May 2018.

Hugh of St. Victor. Erudutionis Didascalicae. Libri Septem. (Liber Secundus. XXI). In Documenta Catholica. http://www.documentacatholicaomnia.eu/02m/1096- 1141,_Hugo_De_S_Victore,_Eruditionis_Didascalicae_Libri_Septem,_MLT.pdf – from Migne Patrologia. Accessed 4 May 2017.

Jaeger, Werner. 1962. *Aristotle: Fundamentals of the History of His Development*. Oxford: Oxford University Press.

Johnston, H. 1967. Liberal Education. In *The New Catholic Encyclopedia*, vol. 8. Washington, DC: CUA Press.

Nightingale, A.W. 2004. *Spectacles of Truth in Classical Greek Philosophy*. Cambridge: Cambridge University Press.

Plato. 1930 (1999 reprint.) *Republic*. 1-5. Trans. Paul Shorey. Ed. Jeffrey Henderson. Loeb Classical Library, Harvard University Press.

———. 1936 (2006 reprint). *Republic*.6-10. Trans. Paul Shorey. Ed. Jeffrey Henderson. Loeb Classical Library, Harvard University Press.

Quintillian. 2001. *Institutio Oratoria/The Orator's Education*. Books 9-10. Trans. Donald Russell. Loeb Classical Library, Harvard University Press.

Sadoleto, Jacopo. 1916 translation. (Nabu reprint.) *Sadoleto on Education. A Translation of the De Pueris Recte Instituendis* Trans. Ernest Trafford Campagnac (translator.) Oxford University Press, 1916 – Nabu reprint.

Seneca, Lucius Annaeus. Moral Essays. Trans. John W. Bassore. 1932 (2001 reprint). Loeb Classical Library, Harvard University Press.

———. *Epistularum Moralem ad Lucillium* LXXXVIII (88) At the Latin Library. http://www.thelatinlibrary.com/sen/seneca.ep11-13.shtml. Accessed 5 May 2018.

Chapter 3
The Political and the Theoretic Life – The Challenge of Socrates

3.1 Nietzsche Against the "Theoretical Man"

This ideal of the theoretic life finds an epitome in a single great historic personality – that of Socrates.[1] As Friedrich Nietzsche writes in *The Birth of Tragedy:*

> To show that even Socrates deserves the dignity of this kind of leading position, one need only recognize in him the archetype of a form of existence unknown before him, the archetype of *theoretical man*...[2]

Nietzsche here recognizes Socrates as an archetypal figure, indeed a "vortex" of world history. Yet he ultimately rejects him as "the mystagogue of science",[3] the figure who submerged the true Greek essence contained in the tragic aesthetic by the theoretical approach which seduced and laid hold of European civilization for over two millennia (Fig. 3.1).

For Nietzsche the modern culture of the Enlightenment with its emphasis on science is really the last, decadent phase of Socratic rationalism, symbolized for him by the figure of Goethe's *Faust*:

> How incomprehensible the true Greek must find *Faust*, the modern man of culture, although he is inherently understandable – Faust, who storms unsatisfied through all the faculties, who has devoted himself to magic and the devil out of the drive for knowledge; we have only to compare him with Socrates to realize that modern man is beginning to realize the limits of the Socratic lust for knowledge...[4]

---

[1] A somewhat different form of this chapter was published online in the Italian cultural web journal Samgha as "The Challenge of Socrates: A Reflection on Philosophy and the Polis" https://samgha.me/2015/02/06/the-challenge-of-socrates-a-reflection-on-philosophy-and-the-polis/ (Accessed September 1, 2018).

[2] Friedrich Nietzsche. *The Birth of Tragedy*. Raymond Geuss and Ronald Speirs (eds.) (Cambridge University Press, 2010): 72.

[3] Ibid. 73.

[4] Ibid. 86.

Fig. 3.1 *The Death of Socrates* by Jacques Louis David (1787) contributed by Everett -Art –www.shutterstock.com

But is Faustianism truly the decadent book end of a Socratic culture? Or is Faust rather the great *rival* of Socrates?

With a moment's reflection we realize that while Faust sacrifices his spiritual good for worldly mastery, Socrates bids the Athenians abandon their concern for power and riches and focus on the perfection of their souls. Still it is no coincidence that so many thinkers (Nietzsche, Spengler, and Berdyaev) see in Faust the very *archetype* of modern civilization with its closure to spiritual concerns, its materialism, and its pursuit of knowledge as an instrument of power. Faustian civilization – that which converts *science into magical power over things* in the form of technology is better understood as *Baconian* civilization. Nietzsche's assumption of a simple continuity between Socratic and modern Enlightenment rationalism ignores the great chasm which exists between them. Nietzsche seems to manifest a general romantic aversion to rationalism (found in other German thinkers like Heidegger) as such, which seeks escape from rationalist modernity through a return to pre-Socratic Greek thought. But there are different streams of rationalism which Nietzsche conflates. Modern rationalism is *scientific* rationalism, and "science" to us means the empirical investigation of physical nature.

Was this the central concern of Socrates? Xenophon provides us with a witness on this point:

> He did not even discuss that topic so favoured by other talkers, "the Nature of the Universe": and avoided speculation on the so called "Cosmos" of the Professors, how it works, and on

the laws that govern the phenomena of the heavens: indeed he would argue that to trouble one's mind with such problems is sheer folly.[5]

Socrates is a rationalist. But the great inquiry of reason he undertakes is not focused on understanding and utilizing the forces of nature. To Socrates it is another inquiry and another passion which occupies his life – and it is one which our modern scientism wholly rejects as outside the domain of knowledge. This is the problem of *the good life*. In particular the central concerns of the modern technological-economic ideal bear little on the great Socratic concern – the care of the soul.

It is *this* concern which ultimately leads to his death. The clash between Socrates and the city of Athens which leads to his trial and death raises in the most dramatic possible form the fundamental question of the relationship between philosophy and politics. What is the nature of this conflict? How was it that Socrates who sought true virtue and wisdom above all else was perceived as a mortal threat to Athenian political life and forced to pay with his life?

3.2 The Political Nature of the Ethical Quest

In point of fact the *conflict* between Socrates and the Polis is explicable only in terms of the fundamental *kinship* between Socratic philosophy and the concerns of politics. Socrates did not make the distinction one finds fully developed in Aristotle between the theoretic life (i.e. the philosophical life) and the political life[6] This contrast with Aristotle is instructive. For him this distinction is rooted in the difference between the moral and theoretical virtues.[7] The moral virtues being active are proper to the political life while the intellectual virtues are proper to the theoretic life. While both the active and contemplative forms of life have their own goodness, Aristotle will argue for the superiority of the theoretic life on the grounds that the intellect is the distinguishing and highest function of man. Moreover, while active works are chiefly good for further things beyond themselves, leisured pursuits have their own proper excellence.

---

[5] Xenophon. *Memorabilia* 1.11 http://www.perseus.tufts.edu/hopper/text?doc=Perseus%3Atext%3A1999.01.0208%3Abook%3D1%3Achapter%3D1%3Asection%3D11 (Accessed 4/3/2015).

[6] Aristotle. *Nicomachean Ethics*. (Loeb Classical Library. Harvard University Press, 1999). H. Rackham trans –cf. the discussion in Book X, 1177aff. I am using Loeb Translations for all the classical texts. These including of Aristotle also the *Politics* (H. Rackham trans., first print in 1932) and of Plato the *Protagoras* (W.R.M. Lamb trans, first print 1924), *Meno* (same volume), *The Apology* (trans. Harold North Fowler, first print 1914), the *Phaedrus* (same volume.) and the *Gorgias* (W.R.M. Lamb first print 1925).

Subsequent references will simply give the text, chapter and Bekker number for Aristotle or Stephanus number for Plato.

[7] Ibid. Book II, 1103.

Had Socrates understood philosophy in this way it might have been possible to virtuously avoid the conflict with the Polis by retreat into a life of solitary speculative activity. But the famed Socratic paradox that "virtue is knowledge" means that for Socrates that philosophy is centrally focused on a question central to politics - to the problem of virtue or moral excellence (ἀρετή). Socrates as we know above all things seeks the knowledge of the good life and to admonish others to seek the same:

> Most excellent man, are you who are a citizen of Athens, the greatest of cities and the most famous for wisdom and power, not ashamed to care the acquisition of wealth, and for reputation and honour, when you neither care nor take thought for wisdom, and truth and the perfection of your soul?[8]

For Socrates the intellectual quest for "wisdom and truth" is part and parcel of the effort at the perfection of the soul, for it is by *knowledge* of the Good that one becomes good. This point is made more explicit in the *Protagoras* where Socrates argues that "…no one willingly goes after evil or what he thinks to be evil in preference to the good."[9] Just as evil is a consequence of and basically identical with ignorance, so is virtue logically identical with wisdom. The philosophical life which seeks the *knowledge* of the Good and the ethical life which *acts* according to the Good are the same for Socrates. This mean that for the philosophical life will also be one concerned with the central question of the political.

But why is the ethical life inherently a *political* life? This assumption is perhaps not as self-evident in our time. Our modern liberalism after all tends to separate the public from the private spheres. Within this framework we may think of the pursuit of the "knowledge of the good" as something relevant for personal and individual life, while politics as concerned with "public matters" – taxation, benefits, foreign policy, rights protection, etc.… But does liberalism actually succeed in evading the question of the Good as a political question? As Leo Strauss pointed out the liberal quest for the "open society" where every individual is simply left free within some broad limits to pursue the good however they understand it, does not succeed in evading the question of the good society (i.e. the central question of classical political philosophy), but is in fact merely a particular *definition* of "the good society."[10]

At all events the notion that emerged in modernity in which the individual is primary, and the Polis secondary was profoundly alien to the Greek mind. Aristotle who more than any philosopher before him affirms the superiority of the theoretic to the political life, nonetheless sees only the political life – i.e. the life lived within the context of the polis – as properly human.

---

[8] Plato. *The Apology*. 29e–30a.

[9] Plato. *Protagoras*.358d.

[10] Cf. Leo Strauss's First Lecture on Meno (Spring Semester 1966) https://leostrausscenter.uchicago.edu/sites/default/files/courses/01%20Plato%27s%20Meno%201%20-%201966-03-29.mp3 (Accessed January 3, 2015). For more on Strauss's views of modern liberalism cf. "Relativism" in Thomas Pangle (ed.) *Classical Political Rationalism*. (Chicago: University of Chicago Press, 1989): 13–26.

If man is by nature social and political, it follows that the good life must be cultivated within the cooperative context of the political community – indeed the ethical activity of man is the chief purpose of politics – as Aristotle declares that "…the Good of man must be the end of politics"[11] and that "the state was formed not only for the sake of life but rather for the good life…".[12]

Socrates in the same manner entertained no thought that the pursuit of the Good would be a merely private, individual affair without implication for the community. The reason is that "…from virtue…come all other good things to man, both to the individual and to the state."[13] Politics after all has ethical implications - knowledge of the Good is as requisite for man's communal as well as individual life.

3.3 The Good in Contention

If both politics and philosophy have the same end – the Good – then how does one account for the clash of the philosophical life represented by Socrates on the one side and the Athenian Polis on the other? The answer must evidently lie in differing conceptions of the Good. For even the city which seems to disclaim the Good entirely and adopts a purely "realist" notion of power politics, nonetheless has a conception of the Good – namely an identification of the Good with power. The issue at stake is whether the particular conception of the Good defended by the Polis can be sustained under the weight of Socratic questioning, or will it be shown to be only an *apparent* and not a *true* Good?

Socratic inquiry aims to discover the answer to the question "what is the good life?" by subjecting all claims concerning wisdom and virtue to the scrutiny of reason, and if they are found wanting to reveal our ignorance in order that the quest may progress. When however, the claimant is the political community itself, philosophy is then seen to potentially represent a force which is subversive of the inherited traditions which provide the moral foundations of the society. "Custom is lord of all" said the Greek poet Pindar, in a phrase famously taken up by Herodotus.[14] The pre-philosophical social conception of the good is typically tied to ancestral, inherited traditions which philosophy calls into question.[15]

That nearly all human societies intuit that *mere* custom or tradition is a self-sufficient justification for its obligatory force– for how does one reason from the fact that something *has* been done to the belief that it *ought* to be done? In general, religious sanctification therefore stands behind custom to provide it with obligatory force. When tradition is associated with the will of God (or the gods) it acquires a

---

[11] Aristotle. *Nicomachean Ethics*. I.ii.7–8.
[12] Aristotle. *Politics*. III. 1280a.
[13] Plato. *The Apology*. 30b.
[14] Herodotus, *Histories*. 3:38. Pindar's original poem is lost.
[15] For a discussion of the Good as the ancestral custom cf. Leo Strauss. *Natural Right and History*. Especially chap. III (University of Chicago Press, Copyright 1953).

special legitimacy. We might then be tempted to see in Socrates a kind of anticipation of the modern rationalists of the eighteenth century Enlightenment; a religious skeptic like Voltaire and Diderot who demands that the claims of religion subject themselves to the critical scrutiny of human reason as part of a grand project of social reform.

In fact, however, the approach of Socratic inquiry toward religion was entirely different. Though Socratic rationalism is broader then Enlightenment rationalism in its concern for capacious philosophical questions, it is also more marked by a sense of humility and intellectual finitude which contrasts with the spirit of modern religious skepticism. When Socrates is asked by Phaedrus if he believes in the myth of Boreas and Oreithyia he responds:

> If anyone disbelieves in these [myths], and with a rustic sort of wisdom, undertakes to explain to explain each in accordance with probability, he will need a great deal of leisure. But I have no leisure for them at all; and the reason my friend is this: I am not yet able, as the Delphic inscription has it, to know myself... so I dismiss these matters and accepting the customary belief about them, as I was saying just now, I investigate not these things, but myself...[16]

In short Socrates' acute awareness of human ignorance and limitations lead him to defer to traditional religious beliefs on matters which he finds himself unable to know or investigate.[17] Socrates also clearly has a religious element in his character as witness by credence he gives to the Delphic oracle which sets him on his philosophical mission, and his belief in a divine Spirit (δαίμων) which guides his moral decisions.

For Socrates however, knowledge of one's ignorance is not valuable merely for the sake of a lazy humility but as a spur to wonder and inquiry. If the individual - or the Polis itself -believes itself to already know the Good what motive will it have to seek out the truth of the matter? As is recorded in the *Meno* with reference to the point at which the slave-boy becomes aware of his ignorance and is perplexed by the mathematical problem:

> Soc: Now by causing him to doubt and giving him the torpedo's shock, have we done him any harm?
> Men: I think not.
> Soc: And we have certainly given him some assistance, it would seem, toward finding out the truth of the matter: for now, he will push on in the search gladly, as lacking knowledge...[18]

Socrates thus has a ready defense against the position that philosophical inquiry is destructive to the polis. He would this deny the claim that the political order would be harmed by rational examination into the foundations of its inherited moral code. For if it should turn out after examination to rest on true foundations confidence in its moral code will be immeasurably strengthened by the transformation of

---

[16] Plato. *Phaedrus.* 229e–230a. (My brackets).

[17] For more on Socrates' religious views see W.K.C. Guthrie. *Socrates.* (London: Cambridge University Press, 1971):155ff.

[18] Plato. *Meno.* 84d.

unreflective opinion concerning the Good into knowledge. But if it should be shown to rest on false foundations then it will be rescued from an ignorance concerning the Good from which it could not otherwise have escaped, and it will be spurred onwards to seek the truth concerning the Good.

This Socratic position however rests on at least two assumptions. First, that human reason is in fact competent to discern the truth concerning the Good – a view which balances Socratic humility. Secondly it presumes that there *is* a truth concerning the Good. To discover this would require scrutinizing the claims of the polis concerning the nature of the political Good. But what was this prevailing conception against which Socrates felt he needed to contend?

3.4 The Rhetor-Statesman or the Philosopher-Statesman?

In Athens, as a deliberative democracy, power was acquired most of all through the art of persuasion – rhetoric, and the statesman was above all things a rhetor. This meant a premium was placed on the skills the Sophists taught – the art of persuasive rhetoric. That the Sophists were central to the development of the Athenian ideal of education is beyond question.[19] To be properly trained in rhetoric required a broad education in the science of argumentation (dialectic) and the beauty of language (poetry). It required also a knowledge of history, politics, and cultures so that one could speak to any audience knowledgably about any theme. Also needed would be a knowledge of human passions and how they are aroused and assuaged. Even the natural sciences and mathematics were not neglected by the Sophists in their education.

The Socratic questioning of this Sophistic education ideal related to the *end* of this education? What is the *value* of rhetoric? Gorgias seeks to impress Socrates by noting that this art "…in itself comprises practically all powers at once!"[20] He provides examples of how with the power bestowed on rhetoric it can do what all other skills can do – for it can for example persuade the patient better than the doctor, and the same with all other skills "So great, so strange is the power of this art."[21] His student Polus awed by the limitless power to which rhetoric seems to give access:

> Are they [the orators] not like the despots, in putting to death anyone they please, and depriving anyone of his property and expelling them from cities as they may think fit?[22]

The presumption of course which Socrates will challenge is that power, and not virtue is the aim of politics.[23] Socrates defends his position that "…to do wrong is

---

[19] Werner Jaeger. *Paideia: The Ideal of Greek Culture, Vol. I* (Oxford University Press, 1962):298ff.
[20] Plato. *Gorgias*, 456a.
[21] Ibid. *Gorgias*, 456c.
[22] Ibid, 466c. (My brackets).
[23] For an excellent discussion of this and many other aspects of the *Gorgias* see Werner Jaeger. *Paideia: The Ideals of Greek Culture*. Vol II. (New York: Oxford University Press, 1943):126–159. Hereafter Jaeger. *Paideia* II.

the greatest of evils"[24] by appealing to the hierarchy of property, body, and soul with the latter the highest. Injustice which is to the soul what disease is to the body is the greatest of evils, just as justice is the greatest good, and so those who make themselves unjust inflict the greatest evil on themselves. It is worse clearly for Socrates, to do that to suffer wrong.[25] If this is so then by analogy the philosophy is like the medicine of the soul, since it brings the knowledge which cures the soul - it of its disease – injustice. But when then is rhetoric?

One is now in a position to understand Socrates' playful analogy between rhetoric and cooking which brings back into focus the central distinction between the pleasant and the Good. The rhetor is to the soul what the cook is to the body – one who brings pleasure without necessarily bringing health. Socrates's opinion of rhetoric is already made clear when he says that "I sum up its substance in the name *flattery.*[26] Just as the cook aims to *please* the body without being necessarily concerned with its health, so the sophist or rhetor gives pleasure to the hearers without necessarily seeking the true good of the soul. The debate therefore is about whether pleasure and the Good are identical or can be distinguished. This comes out in the discussion with Callicles will later defend the identity of the pleasant and the Good.[27] However, under the relentless cross examination of Socrates, Callicles is compelled reluctantly to admit the distinction.[28] This establishes the possibility that the pleasure produced by rhetoric can be only an apparent good.

The failure to distinguish pleasure from the Good is at the root of the whole conflict between Socrates and the Athenian polity. The people in the generality will far prefer the pleasing flattery of the rhetor to the philosopher's instruction, just as they will prefer the cook who brings the body pleasure even at the cost of health, to the doctor who brings the body health even at the cost of some pain. Unable to discern the distinction between the apparent Good brought by the rhetor and the true Good sought by the philosopher, they will resist the latter – or even seek to destroy him. As Socrates prophecies ". "I shall be like a doctor tried by a bench of children on a charge brought by a cook."[29]

If, however one then accepts the Socratic argument – that the good of the soul (virtue) is the true aim of politics rather than power and pleasure, it follows that the polis will require someone to pursue the knowledge of the Good. If that is true it is the philosopher who seeks this knowledge and not the rhetor who is the truly political man. As Socrates notes to his most formidable opponent Callicles:

> I think I am one of the few, not to say the only one, in Athens who attempts the true art of statesmanship, and the only man of the present time who manages affairs of state:... the

---

[24] Plato. Gorgias,469b.

[25] Ibid.469C – see also 477c and following.

[26] Ibid. 463b-c.

[27] Ibid.495a.

[28] Ibid. 500e.

[29] Plato. *Gorgias*. 521e.

speeches I make from time to time are not aimed at gratification, but at the best instead of what is most pleasant...[30]

So far then from accepting the Aristotelian distinction between the political and the philosophical life, Socrates position turns on the claim that the only *truly* political life *is* the philosophical life.

Socrates thus raises the problem concerning the tension between the philosophical which leads to virtue and knowledge and the political life which leads to power. This idea was of the utmost influence on the whole classical world in posing a fundamental challenge of how philosophy and politics could be reconciled. It was answered in three distinct ways. First, there was Plato who following the Socratic idea famously conceived of the idea of the philosopher-kings in *The Republic* – thus conjoining the philosophical and political life in the way already hinted at in the *Gorgias*. This proposal however has proven rarely practicable since philosophers are rarely rulers. Second was the answer of the schools of rhetoric (e.g. Isocrates) who argued for a corresponding ideal – the rhetor-statesmen who is nonetheless infused by that knowledge of virtue imparted by moral philosophy. This idea proved to be of enormous significance not only in the Greek but also in the Roman rhetorical tradition represented by Cicero and Quintilian. Both of these figures argue for the union of eloquence with virtue and wisdom, or in short of moral philosophy with rhetoric.[31] This ideal of Ciceronian humanism proved enormously influential historically being forcefully revived during the Italian Renaissance.[32] In spite of the fertility of this idea, as a conception of the highest form of life it sidelines the importance of the speculative activity of philosophy (metaphysics) in favor of the practical. The final resolution of the problem was that of Aristotle, which involved giving the political life its due place. But establishing the autonomy of the theoretic life.

References

Aristotle. 1926. (1999 reprint). *Nicomachean Ethics*. Trans. H. Rackham, 1999. Loeb Classical Library, Harvard University Press.
Cicero. *De Oratore*. 1948. Trans. E.W. Sutton. Loeb Classical Library, Harvard University Press.
Herodotus. 2006 (reprint). *The Persian Wars*. Books III-IV. A.D. Godley (trans.) Loeb Classical Library, Harvard University Press.
Jaeger, Werner. 1962 (reprint). *Paideia: The Ideal of Greek Culture. Volume I*. New York: Oxford University Press.
___. 1963 (reprint). *Paideia: The Ideals of Greek Culture. Volume II*. New York: Oxford University Press.

---

[30] Ibid. 521d.
[31] Some examples would be Cicero's *De Oratore* I xv, and Quintillian's *De Institutio Oratoria* XII.2.
[32] Quentin Skinner. *The Foundations of Modern Political Thought*. Vol 1(Cambridge University Press, 2010): 87.

Nietzsche. Friedrich. 1999. *The Birth of Tragedy.* Raymond Geuss and Ronald Speirs (eds.). Cambridge University Press, 2010.

Plato. 1924 (first printing). *Laches. Protagoras. Meno. Euthydemus.* Trans. W.R.M. Lamb. Loeb Classical Library, Harvard University Press.

———. 1996 (reprint). *Lysis. Symposium. Gorgias.* Trans. W.R.M. Lamb. Loeb Classical Library, Harvard University Press.

———. 2001 (reprint). *Euthyphro. Apology. Crito. Phaedo. Phaedrus.* Jeffrey Henderson (ed.). Loeb Classical Library, Harvard University Press.

Quintillian.1020–1922 *De Institutio Oratoria.* At University of Chicago. http://penelope.uchicago.edu/Thayer/E/Roman/Texts/Quintilian/Institutio_Oratoria/home.html. Accessed May 2018.

Rosenthal Pubul, Alexander. 2015. "The Challenge of Socrates: A Reflection on Philosophy and the Polis.". https://samgha.me/2015/02/06/the-challenge-of-socrates-a-reflection-on-philosophy-and-the-polis/ Accessed 1 Sept 2018.

Skinner, Quentin. 2010. *The Foundations of Modern Political Thought. Volume 1.* The Renaissance. Cambridge University Press.

Strauss, Leo. *Natural Right and History.* 1953 (1999 reprint). Leo Strauss. University of Chicago Press.

———. *Natural Right and History.* 1966. Leo Strauss's First Lecture on Meno (Spring Semester 1966). University of Chicago. https://leostrausscenter.uchicago.edu/sites/default/files/courses/01%20Plato%27s%20Meno%201%20-%201966-03-29.mp3. Accessed 3 Jan 2015.

———. 1989. "Relativism." In *Classical Political Rationalism,* ed. Thomas Pangle. Chicago: University of Chicago Press.

Xenophon. 1923 (2013 revision). *Memorabilia. Oeconomicus. Symposium. Apology.* Trans. E.C. Marchant. O.J. Todd. Loeb Classical Library, Harvard University Press.

Chapter 4
Aristotelian Teleology: The Bridge Between Natural Philosophy and the Problem of "The Good Life"

4.1 Aristotle and Plato

Though Socrates embodies the theoretic life perhaps more fully than any other philosopher, we have seen that his concept of the theoretic life is inextricably bound with an idea of political-ethical virtue. Indeed, Socrates will defend philosophy as precisely the highest form of political-ethical activity because it seeks knowledge and not merely opinion concerned the Good.

Aristotle therefore assumes for us the most fundamental importance. It is only with Aristotle that the theoretic life first acquires its full *autonomy* as a form of life distinct from others including the political life. Insofar as "ethics" is considered capaciously as pertaining to the good life it remains relevant – for the theoretic life is still the very highest form of human life. Yet wisdom appears as its own end with its proper excellence, and not (merely)as a requirement for the practice of the moral virtues. It is this which most essentially distinguishes the Aristotelian from the Platonic understanding of it. For Plato because virtue is identified with wisdom there is no essential distinction made between the practical and the theoretical pursuit of the Good. Not only is philosopher is identified as the true statesman, but a central theme in the *Republic is* that the more the philosopher and statesman coincide in the ruler, the more virtuous the city will be.[1]

In this chapter we will discuss Aristotle's argument for the theoretic life and its distinction from – and superiority to – other forms of life including the political life. In order to properly understand this argument, it will be important to brief exposition of the importance of the premises of Aristotle's natural teleology in his overall argument.

---
[1] Plato. *Republic*. 473d.

4.2 Natural Teleology

Natural philosophy and ethics/politics have two generally distinct lineages in Greek thought, the former coming from the Pre-Socratic Ionian natural philosophers (the φυσιολόγοι), while Socrates is generally conceded to be the father of Greek political and moral philosophy.

It was the supreme achievement of Aristotle to integrate these two streams of philosophy into a common system. On the one hand one his works in practical philosophy (e.g. the *Nicomachean* and *Eudemian Ethics* and the *Politics* focused on the idea of the Good in the (for him) closely related disciplines of ethics and politics. And then are his works in theoretical philosophy in which we may place his natural philosophy (e.g. the *Physics*) including physics and cosmology as well at the highest level of universality metaphysics and ontology (e.g. the *Metaphysics*.) At first sight these two dimensions of the Aristotelian corpus might seem entirely disparate from each other. And yet in a real sense the idea of τέλος – purpose and end – can be understood as the lynch pin of the Aristotelian philosophic system as a whole, connecting Aristotelian natural philosophy and metaphysics, and Aristotelian ethics and politics. The Aristotelian cosmos is a hierarchy of beings each ordered to its proper end or good which perfects and fulfills its nature (as for instance a natural intrinsic finality directs acorns to become trees). It is because beings have ends in nature, and because man is part of nature that man can be considered to have an end. This furnishes a basis for Aristotle's ethical system in that the good life will be the one which most fully fulfills the natural end or function of man. It is toward this end that the whole of human moral and political life ought to be directed. *Teleology thus forms the link in Aristotelian though between Being and nature on the one side and the Good on the other, which may be considered therefore as a final cause with respect to man.*

This leads to two questions first what the character of nature is (is it purposeful?) and secondly what man's place within nature is.

The Greek idea of "nature" (φύσις) is one characteristic and central to Greek philosophy from its very origin, the first Greek philosophers in Ionia being known as "physicists (φυσιολόγοι). Figures like Thales, Anaximander, and Anaximenes achieves the critical breakthrough to philosophy by breaking from mythical forms of thought to analyze nature. They sought to determine for instance which was the fundamental element from which the others arose. Later Greek philosophers the Eleatics, Heraclitus, Empedocles, Anaxagoras, and the Atomists were absorbed with such problems as explaining the nature of change and whether the universe was one or many. For the Atomists like Democritus nature seems to be a process of unchanging atoms producing change by their constant motion and rearrangements but without any purposeful ends. In this way he is a forerunner of the basically mechanistic and non-teleological theories of early modernity as found in Bacon,

Aristotle who returned so forcefully to the problems of natural philosophy gave vital importance to the idea of the final cause – the purpose within natural forms and processes. Nature is filled with purposeful activities – one may think for example of the distinctive functions of the bodily organs – the adaptation of the eye for sight or the teeth for mastication. Or one might think of developmental processes as the motion of an acorn to become a tree, or an embryo to become a bird or mammal. Aristotle would see it as absurd to ascribe such processes to mere chance or fortune.

> ...when the desirable result is effected invariably or normally, it is not an incidental or chance occurrence; and in the course of Nature the result always is achieved either invariably or normally, if nothing hinders...that Nature is a cause, then, and a goal-directed cause, is above dispute.[2]

The pillar of the Aristotelian conception of nature and causality lies then in the idea of the final cause or purpose. According to this there is a motion and potency within the things of nature toward a purpose or end which perfects them. Aristotle himself gives a special primacy to the final cause and when he remarks that the physicist must regard both the material and final cause, but the former is itself caused by – and hence subordinate to – the latter:

> ...though the physicist has to deal with both material and purpose, he is more deeply concerned with the latter; for purpose directs the moving causes that act upon the material, not the reverse. It is the goal that determines the purpose, and the principle of causation is derived from the definition and rationale of the end, in Nature, just as much as in artificial constructions.[3]

He earlier[4] provides the useful example of the saw whose definition is supplied by its end (sawing) and who material (iron) must be adapted to this end rather than the reverse. In the history of philosophy, he provides at the outset of the *Metaphysics* he presents final causality implicitly as his own culmination of the process of philosophical thought concerning causes. In his schema the Ionian Natural philosophers like Thales conceived only the material cause.[5] Later philosophers like Anaxagoras, Empedocles, and the Atomists conceived also the efficient cause, which the Platonists ignored in favor of the Formal cause. The Final Cause however he says was only inadequately perceived.[6]

4.3 Man Within Nature – Is There a Natural Human Telos?

Socrates is credited by Cicero with bringing philosophy from the heavens into the world of men. This raised a new question. What is the place of man within the order of nature? Can nature provide a key or model to the good life? The task of employing nature as the source of ethical norms begins already with Plato. In the *Republic*

---

[2] Aristotle. *Physics*. II. VIII (199b).
[3] Aristotle. *Physics*. II.ix (200b).
[4] Ibid. 200a.
[5] Aristotle *Metaphysics*. A 983b.
[6] Ibid. 988b5.

he speaks of the virtuous life in relation to the soul on the model of health in relation to the body stating that "…to act unjustly and be unjust and in turn to act justly… these are in the soul what the healthful and diseaseful are in the body…".[7]

The state of health for Plato is one of due order and balance – the proper subordination of lower elements in the body to the higher. The same therefore will be true of the soul – in the just soul the higher element (reason) will control the lower element (the appetites.) Justice and virtue then will be a state or activity according to nature (κατά φυσιν) while vice and injustice will be a state of soul contrary to nature (παρά φύσιν) such that "virtue will be…a kind of health and beauty and good condition of the soul."[8]

Three essential things can be derived from Plato's discussion. The first that that nature is not confined to the "material" world – the human soul also belongs to nature. The second is that for Plato nature is the source of the moral norm – to live well is to live according to nature.[9] The third is that Plato already has a kind of "proto-teleology" in his ethical theory; there is a natural good to which each part of human nature – body and soul alike – ought to conform for its perfection.

With Aristotle the teleological conception of nature is fully realized. Nature for Aristotle is characterized fundamentally by a distinction and hierarchy of functions – an order of "ruling and being ruled".[10] There is the hierarchy of plants, animals, and humans, and within man of reason over the appetites and of the soul over the body. This perforce raises the question of what precisely the ultimate End of man is – a question which is equal to asking what is the supreme Good at which human action properly ordered aims?

4.4 The Supreme Good

The distinction between the good in itself and the useful raises a further question. Some goods are useful in the sense that they are sought for the sake of a further good. But can *all* goods be of this nature? Can we imagine an infinite set of goods to be pursued, each of which desired for the sake of some further good? Or are we required to imagine that there is an end to the chain – some supreme Good sought for its own sake? The question is significant for a basic contention that arises in modern philosophy is that there is no ultimate or supreme good at which all human action aims (or ought to aim.) Rather it is argued there is simply an unending, endlessly variegated series of objects to human desires. This contention is pregnant with the most important consequences for ethical and political philosophy. For if there is no supreme or highest Good, then there is in effect no way to determine which among a range of objects of desire are better or worse. One is left in that case

---

[7] Republic. 444d.
[8] From 444e – I switched to small letter "v".
[9] See Werner Jaeger. *Paideia* II.:322.
[10] Aristotle. *Politics*. (1254a).

with a species of relativism or at the least value pluralism – which is today so common.

The greatest among the ancient philosophers argue for the existence of the highest Good. Plato provides a vision of *the* Good as a transcendent Reality beyond even Being.[11] Aristotle however rejects this conception, for such a Good will not be attainable by man and so is not a practical object of human desire.[12] Aristotle does however think there must be *some* supreme Good desired for its own sake, for if everything were desired only for the sake of some further good there would be no ultimate account of human action at all:

> If therefore among the ends at which our actions aim there be one which we wish for its own sake, while we wish the others only for the sake of this, and if we do not choose everything for the sake of something else (which would obviously result in a process *ad infinitum*, so that all desire would be futile and vain), it is clear that this one ultimate End must be the Good and the Supreme Good.[13]

Aristotle will argue that happiness (εὐδαιμονία) is such a good, for no one does anything unless he thinks it will advance his happiness (or diminish his unhappiness), and no one reasonably thinks of happiness as a good serving some further end. But this does not according to Aristotle address the issue of what good happiness consists in. The question has been partially addressed at least by way of negation. All things deemed worthy of desire are either useful, pleasant, or good. But we established that the useful cannot be the *supreme* Good since things are useful only in relation to some further good.

4.5 The Different Types of Life and the Question of the Best Life

One of the most valuable innovations of Greek thought, perhaps commenced by Socrates and Plato is the idea of the βίος – a type of life considered as a "clear and comprehensible unity, a deliberately shaped life pattern"[14] and the "expression of a particular ethos."[15] Since it embraces the unity of a human life across time it provides a complete and all-embracing concept for the ordering of the whole of life. This idea is rather easily assimilated to the Aristotle's concept of the supreme good in a teleological sense. Each person seeks happiness principally through the possession of some good which regard as of supreme value or worth. In fact, the three

---

[11] Plato. *Republic,* 509b.

[12] Aristotle. *Nicomachean Ethics* I. Vi.13. Interestingly later Christian thinkers like St. Thomas Aquinas will in a sense combine the Platonic and Aristotelian perspectives by conceiving of the transcendental Good as attainable but only in the order of grace not nature. Hence the *visio beatifica* – the vision of the divine Essence.

[13] Aristotle. *Nic. Ethics*. I.1.2.

[14] Jaeger. *Paideia* II.46.

[15] Ibid. 349.

types of life which Aristotle chiefly considers as candidates for the best – the Life of Pleasure or Enjoyment (βίος ἀπολαυστικός), the political life, and the theoretic life,[16] already appeared in Plato's *Republic* as the three classes of persons in the city who represent the three elements of the soul and consequently the three chief pursuits of man.[17] The majority of mankind for both Plato and Aristotle equate happiness with *pleasure* and are ruled then by their appetites and desires, particularly those of a physical nature. This Aristotle terms the life of enjoyment. A smaller number whom Aristotle terms "men of refinement" equate happiness instead with *honor*. Many indeed (one might think of soldiers) are willing to endure hardship and pain for the sake of honor. This kind of active life of noble deeds has of course a long lineage as an ideal in Greek thought going to Homer and was widely held as the best and most ideal form of life. In the classical context of the polis, Aristotle terms this kind of life the political life (βίος πολιτικός) because it is most concerned with moral virtues exercised for others in the context of the public life of the community. And finally, there are the philosophers who see the pursuit of *wisdom* as the highest good and happiness contemplative activity. This is the kind of life Aristotle terms the theoretic life (βίος θεωρητικός).

No other form of life beyond these three is seriously considered by Aristotle as a viable "candidate" for the noblest and best type of human life. One might be tempted to consider money such a supreme good since so many occupy so much time in its zealous pursuit. Yet the "Life of Money-Making" is quickly rejected as a viable competitor to the other three on the grounds that:

> ...clearly wealth is not the Good we are in search of, for it is good only as being useful, a means to something else. On this score indeed one might conceive the ends before mentioned [pleasure, honor, and wisdom] to have a better claim, for they are approved for their own sakes.[18]

Money in short as a medium of exchange by which one can acquire other goods, and thus cannot be the supreme Good. People as a rule desire money for other goods which it can get them, such as the necessities which all require as a condition of pursuing their ends, or pleasures and enjoyments. Though it is certainly possible to use money in the pursuit of honor or wisdom as well. But in none of these cases is it more than a useful means to an end. Its utility again shows to be a good of a subordinate order.

Once it is apprehended that there is a supreme Good of human life, and that by definition this supreme Good cannot belong to the category of "the useful" there are then three viable types of life – the Life of Pleasure or Enjoyment, the Political Life, and the Theoretic Life, corresponding to the three goods which are sought for their own sake – pleasure, honor and wisdom. Granting that human beings will disagree vehemently over which of these is best, it remains for Aristotle to show which of

---

[16] Aristotle. *Nicomachean Ethics*. I. Iv.5.
[17] Plato. *Republic* Book IV 580dff and see the discussion in Jaeger. *Paideia* II. 349.
[18] Aristotle. *Nicomachean Ethics*. I.v.8. (My brackets)

4.6 The Life of Enjoyment (βίος ἀπολαυστικός)

these is in *truth* the noblest and best form of human life, in the sense of realizing the highest excellence of human nature. This will require analyzing each in turn.

4.6 The Life of Enjoyment (βίος ἀπολαυστικός)

To ask whether the Life of Enjoyment is the best form of human life is to ask whether pleasure has the character of the supreme good. In others whether pleasure and "the Good" are identical or distinct. The temptation to identify the supreme Good with pleasure – hedonism – has been a recurrent motif in the history of philosophy. The reason for this is perhaps because as Aristotle says:

> ...that thing which is most desirable which we choose not as a means to or for the sake of something else; but such admittedly is pleasure: we never ask a man for what purpose he indulges in pleasure – we assume it to be desirable in itself.[19]

Unlike utility, pleasure then does *seem to* have its own intrinsic attraction. Evidence for the power of hedonism over utilitarianism as a doctrine of the good is shown by the fact that even the modern utilitarian privileges pleasure as the very criterion of utility arguing that since "Nature has placed mankind under the governance of two sovereign masters, pain and pleasure…".[20]

The Pleasure Principle thus furnishes the foundation for the whole Utilitarian theory of ethics, politics, and law. Hedonism in philosophy seems to have been forcefully revived when Hobbes identifies pleasure with the Good.[21] This identification of pleasure with the Good seems central also to the whole ethos of modern consumerism which implicitly and practically privileges pleasure, comfort, entertainment and the endless satisfaction of desires as "the good life." Among the moderns Hobbes identifies pleasure with the sense of good.[22] Among the ancients some or other version of hedonism was credited to Eudoxus, and the Cyrenaics and perhaps most famously Epicurus who argued for the intrinsic goodness of all pleasure the distinctions among them being caused only by the difficulties they entail which may be greater than the pleasures they produce.[23]

Plato recognizes the importance of the problem. In his *Protagoras* he provisionally or hypothetically accepts the argument that all actions which aim at the pursuit of pleasure and the avoidance of pain are good.[24] He uses this to argue that even on this account vices are caused by defects in knowledge – that is measuring the longer term consequences of pain against the more immediate pleasures.[25] However in the

---

[19] Aristotle. *Nicomachean Ethics*. Xii.2
[20] Jeremy Bentham *An Introduction to the Principles of Morals and Legislation*. I.1.
[21] Hobbes. *Leviathan* I. 6.
[22] Hobbes. *Leviathan* I. 6.
[23] Epicurus. *Principal Doctrines*. 8.
[24] Plato. *Protagoras*. e.g. 358b.
[25] Ibid. 357d.

Philebus which is principally devoted this issue he contrasts the life of pleasure to the life of wisdom. Arguing that if one took each in its pure form without admixture of the other no one would consciously choose pleasure without mind.[26] This provides an important foundation for the defense of the theoretic life over a life devoted to pleasure. The question of whether pleasure and the Good are identical is likewise central to the Socratic argument in the *Gorgias*. Socrates identifies the art of rhetoric with a form of flattery which "…cares nothing for what is best, but dangles what is most pleasant for the moment as a bait for folly…"[27] Rhetoric is to true philosophy like cookery in relation to medicine. The former is concerned to delight the body without regard to its good, as the latter seeks the good of the body more than its pleasure. Thus, it appears that not all pleasures are beneficial even where they seem so. Since the excess of certain pleasures can harm the body it is clear that pleasure cannot be identified with the good as such.

Similarly, philosophy seeks the good of the soul, while rhetoric only aims to intoxicate it with apparent goods. Callicles in the dialogue likewise strongly defends the hedonist position. When Socrates asks him to "…tell me whether you say that pleasant and good are the same thing or is there some pleasure which is not good." He responds, "I say they are the same."[28] In the course of the discussion however Socrates enables Callicles to see a distinction in that unrestrained or licentious fulfillment of desires is harmful to the body and the soul and hence "…only those desires which make men better by their satisfaction should be fulfilled, but those which make him worse should not…".[29] Hence it is apparent that the good and the pleasant are distinct and pleasure cannot be the supreme Good since it can co-exist with evil.

Aristotle considers this question with great seriousness and in detail aiming to determine whether pleasure is a good at all, and if a good is the supreme good. His basic foundation of his argument is that pleasures is associated with activities and differ in moral quality according to the activities with which they are associated – "…the pleasure in a good activity is morally good, and that of a bad one morally bad."[30]

From this is clear that while pleasure is not intrinsically evil, pleasure per se cannot be the *supreme* good since it can also be associated with base deeds. Indeed, one can discern whether a man is good or evil from the very things in which he takes pleasure as to whether they are noble or base – a possibility which necessitates the possibility of distinguishing good from pleasurable actions. Pleasure will therefore play a role in the highest and best kind of life and it will involve taking pleasure in the best mode of human activity. However, a life which treats pleasure as the

---

[26] Plato. *Philebus*.21 b-d.
[27] Plato. *Gorgias*. 464D.
[28] Ibid. 495a.
[29] Ibid.503C-D.
[30] Aristotle. *Nic. Ethics*. X.6.

supreme good in itself and so pursues all pleasures without distinction will according to Aristotle be lowly, animalistic and base – in short is "…only a life for cattle."[31]

4.7 The Function Argument

This points to the larger reason why pleasure is for Aristotle not suitable to the supreme good, nor correspondingly ca the life of pleasure be the best form of life. The supreme good of man must be particular to the human form of being. Pleasure as such, and particularly the physical pleasures like food and sex are shared between humans and animals. Consequently, a life which makes its focus the pursuit of such physical pleasures will be an animalistic life rather than a properly human one.

This is related to Aristotle's argument concerning function (ἔργον). The good of anything is related to the right performance of its function which is particular to it – as a good musician is one who plays well, or a good nutcracker is one which cracks a nut well. In the natural order also, Aristotle presumes an order of beings which with its own function which perfect and fulfill their nature. Hence "…the good of man resides in the function [ἔργον]of man."[32]

But how to find the function? The good as well as the function of anything is related to what is particular to it – as a good musician is one who plays well, or a good nutcracker is one which cracks the nut well. The speak of the "function of man" is the same as to find what is particular to man, that is to the human mode of being.

Not only is there a hierarchy of beings in nature – non-living, plant, animal and human – there is a corresponding hierarchy of powers within man nutrition and growth which he shares with plants, sensation which is shared with animals. But reason according to Aristotle is particular to man and distinguishes him from all the other animals. Hence, he concludes concerning the function(ἔργον") that "…the function of man is the active exercise of the soul's faculties in conformity with rational principle…".[33]

The good life then lies in the full development of what is particular to man as a rational animal – the intellectual and moral faculties. This life of reason will moreover be identical to the life of virtue – "…the active exercise of his soul's faculties in conformity with excellence or virtue…".[34]

Reason however is exercised both in active moral virtues where it consists in the ordering and moderating of the passions and actions in accordance with the rational mean.[35] But it is also exercised in purely theoretical activities like philosophy. So, is the highest form of human life to be found in an active life of noble deeds, or in the

---

[31] Aristotle.*Nicomaechean Ethics* I.V.3.
[32] Aristotle. *Nicomachean Ethics*. I. vii.10. (My brackets)
[33] Ibid. I.vii.14.
[34] Ibid.I. vii.15.
[35] Aristotle. Nicomachean Ethics. II.vi.15.

contemplative activity as for instance the philosopher? The first type of human life Aristotle terms "the political life" and the second type he calls "the theoretic life."

At all events these two types of life are as is clear from the function argument the only kinds of life worthy of man. But which of them is the greater remains to be decided. Let us examine each of them in turn.

4.8 The Political Life (βίος πολιτικός)

The original archaic Greek ideal of excellence or virtue (ἀρετή) was an active one. From the heroes of Homer, the Greeks learned the off quoted maxim αἰεν ἀριστεύειν – "ever to excel"[36] by which they meant to achieve excellence and the prize of honor through noble and heroic deeds. A philosophically refined notion of such an active life of moral virtue is found in Aristotle's depiction of the "great souled man" (μεγαλόψυχος)[37] whom he presents as possessing the summit of all the virtues. Indeed, his quality of greatness of soul is like "…a crowning ornament of the virtues: it enhances their greatness, and it cannot exist without them."[38] For his excellence like the Homeric heroes he claims the reward of honor, at least from those worthy to judge it, for "…honour is the prize of virtue and the tribute we pay to the good."[39]

It remains a question why for Aristotle this idea of an active life of moral virtues should be understood as a *political* life. We should not understand here by "political" the narrower modern idea of "state affairs" as if Aristotle was thinking about the life of a functionary in a modern state bureaucracy. Rather he is referring to the "polis", the community of which the Greek free man formed an integral part and participated as citizen. To say such a life of moral virtue is political is simply to say that it is lived within the community. The classical Greeks could hardly understand the life of an individual apart from the polis. The life of active moral virtue is political in the sense one can hardly be generous or just unless one lives among one's fellows and can treat them with generosity and justice. To be "political" is to be social and communal. In his *Politics* Aristotle makes clear that that man's very nature is political, for both the insufficiency of the solitary individual. Indeed, the human being outside the polis or political community is deemed to be somewhat monstrous. For the man who is "…by nature and not by fortune citiless is either low in the scale of humanity or above it (like the 'clanless, lawless, heartless' man reviled by Homer, for he is by nature citiless…".[40]

Further testimony of the "political" nature of man is found in the unique human faculty language and; for if language has no utility without other human beings. By

---

[36] Homer. *Iliad*. VI.208.
[37] Aristotle. Nicomachean Ethics. IV.iii. See also Jaeger. *Paideia* I. 11–14.
[38] Ibid.IV.ii.16.
[39] Ibid.15.
[40] Aristotle. *Politics*. 1253a.

"political" Aristotle means principally "communal" or "social" as the Polis was in essence a self-governing community. So natural is the political-communal life to man to man that the individual outside of the Polis is inhuman either by way of superiority of beastly inferiority:

> ...man is by nature a political animal, and a man that is by nature and not merely by fortune citiless is either low in the scale of humanity of above it...and why man is a political animal in a greater measure than any bee or gregarious animal is clear. For nature we declare, does nothing without purpose; and man alone of the animals possesses speech.[41]

This power of "speech" (λόγον) is moreover proper to man for through speech one can communicate not only concerning pleasure and pain but also concerning good and evil.[42] This point by Aristotle directly ties the social nature of man revealed by language, to the rational and *moral* nature of man. *The life of moral virtue can only be cultivated within the social context of the polis.*

This idea is particularly developed in the last chapter of the *Nicomachean Ethics*. Here Aristotle argues that teaching by itself will be insufficient for many to the end of leading them to virtue:

> For he that lives at the dictates of passion will not hear or understand the reasoning of one who tries to dissuade him...speaking generally, passion seems not to be amenable to reason, but only to force.[43]

Nor is paternal authority within the family or of any other individual sufficient to curb the wayward passions. What is needed is a system of laws which Aristotle defines as "...a rule, emanating from a certain wisdom and intelligence, that has compulsive force."[44] Laws of course presume and require a political order. Since a properly human life is a political life, and to the polis and that it is only in such a social context that the moral virtues can be exercised it would seem natural to assume that the political life would be the highest and most perfect for man. And yet the political life has a competitor in the theoretic life.

References

Aristotle. *Nicomachean Ethics*. 1926 (1999 reprint). *Nicomachean Ethics*. Trans. H. Rackham, 1999. Loeb Classical Library, Harvard University Press.
———. 1929 (1980 reprint). *Physics*. I–IV. Trans. Phillip H. Wicksteed and Francis M. Cornford. Loeb Classical Library, Harvard University Press.
———. Politics. 1932 (2005 reprint). H. Rackham. Loeb Classical Library, Harvard University Press.
Bentham, Jeremy.1781. An Introduction to the Principles of Morals and Legislation. Utiliarianism. com https://www.utilitarianism.com/jeremy-bentham/index.html#one. Accessed May 2018.

---

[41] Aristotle. *Politics*. 1253a.
[42] Ibid.
[43] Aristotle. *Nicomachean Ethics*. X.ix7.
[44] Ibid. X.ix.8.

Epicurus. Principal *Doctrines*. Epicurus.net http://www.epicurus.net/en/principal.html. Accessed May 2018.

Hobbes, Thomas. 1660. The Leviathan. At TTU. http://www.ttu.ee/public/m/mart-murdvee/EconPsy/6/Hobbes_Thomas_1660_The_Leviathan.pdf. Accessed May 2018.

Jaeger, Werner. 1963(reprint). *Paideia: The Ideals of Greek Culture,* Vol II. New York: Oxford University Press.

Plato. 1930 (1999 reprint). *Republic,* 1–5. Trans. Paul Shorey. Ed. Jeffrey Henderson. Loeb Classical Library, Harvard University Press.

———. *Republic.* 6–10. Trans. Paul Shorey. Ed. Jeffrey Henderson. Loeb Classical Library, Harvard University Press.

———. 1996 (reprint). *Lysis. Symposium. Gorgias.* Trans. Harold Fowler and W.R.M. Lamb. Loeb Classical Library, Harvard University Press.

———. 1925 (1995 reprint). *The Statesman. Philebus. Ion.* Trans. Harold Fowler and W.R.M. Lamb. Loeb Classical Library, Harvard University Press.

Chapter 5
The Aristotelian Revolution: The Autonomy of the Theoretic Life and the Dream of Universal Science

5.1 The Theoretic Life (βιος θεωρητικός)

The first question to ask might be – why does Aristotle distinguish the political and theoretic life to begin with at all? We have already seen with Plato how he represents virtue as a form of knowledge, and consequently the philosophical or theoretic life as the highest form of political life. In other words, the theoretic and political life are identified with each other by Plato.

Aristotle's most radical move then relative to the Platonic conception is to establish the *autonomy* of the theoretic life by *contrasting* it with the political life.

Instead of making the philosophical life the perfection of the political life, Aristotle distinguishes the two. The basis of this distinction lies in his position that there are in fact two forms of wisdom– practical wisdom (φρόνησῖς) and theoretical wisdom (σοφία). Practical wisdom, the wisdom needed to undertake correct moral or political actions in concrete circumstances hinges not only on the abstract knowledge of principles but also on knowledge of the particular circumstances. In essence it applies general principles in a practical way to particular circumstances. Wisdom as "practical prudence" then hinges on experience

> Nor is Prudence [φρόνησῖς] a knowledge of general principles only : it must also take account of particular facts, since it is concerned with action, and actions, and action deals with particular things.[1]

Thus, Aristotle notes that:

> ...although the young may be experts in geometry and mathematics and similar branches of knowledge, we do not consider that a young man can have Prudence. The reason is that Prudence includes a knowledge of particular facts, and this is derived from experience which a young man does not possess...[2]

---

[1] Nicomachean Ethics. VI.vii.7. (My brackets)
[2] Ibid. VI.viii.5.

However, we have already seen that Aristotle considered wisdom in the theoretical sense (σοφία) as superior in character to experience, insofar as knowledge of universal principles is a higher form of knowledge and more truly wisdom then experiential knowledge of particular things. Moreover, wisdom in the theoretic sense has more noble and dignified objects since it is concerned with more than merely human affairs but also of the cosmic and divine orders.

> Hence it is clear that Wisdom [σοφία] must be the most perfect of the modes of knowledge…it must be a consummated knowledge of the most exalted objects. For it is absurd to think that Political Science or Prudence [πολιτικὴν ἢ τὴν φρόνησιν] is the loftiest kind of knowledge, inasmuch as man is not the highest thing in the world.[3]

Corresponding then to the distinction between the theoretic and practical forms of wisdom, is the distinction grounded in the division the moral (ἠθικῆς) and the theoretical (διανοητικῆς) virtues.[4] Depending on which virtues are given primacy distinction leads to one between two modes of life – the political life (βίος πολιτικός) focused on moral virtue, and the theoretic life (βίος θεωρητικός) focused on intellectual virtue.

The most significant defense of the superiority of the theoretic life over the political life can be found in Book X of the *Nicomachean Ethics*. This superior nobility of the theoretic life is shown in several ways of which the criteria of leisure hitherto discussed may be the most important. The highest forms of human activity will be those done for their own sake for, which is to say for leisure, as opposed to labor which is done for the sake of some further thing, for "…happiness is thought to involve leisure, for we do business in order that we may have leisure…."[5]

We have already seen that for Aristotle activity which has its own intrinsic goodness ranks above what is merely useful as productive or instrumental to attain some further good. But the political life of practical virtue is of this "un-leisured" kind for its activity aims at something beyond itself:

> …the activity of the politician also is unleisured, and aims at securing something beyond the mere participation in politics – positions of authority and honour, or if the happiness of the politician himself and of his fellow-citizens, this happiness is conceived as something distinct from the political activity…among practical pursuits displaying the virtues, politics, and war stand out pre-eminent in nobility and grandeur, and yet they are unleisured, and directed to some further end, not chosen for their own sakes…[6]

By contrast the theoretic life is "leisured", for while labor aims at producing some result beyond the work itself, leisured activity like that proper to the theoretic life aims at nothing beyond itself for the very activity is in itself a good of the highest excellence:

> …contemplation may be held to be the only activity which is loved for its own sake; it produces no result beyond the actual act of contemplation …the activity of the intellect is

---

[3] Ibid. VI.vii.2–4. (My brackets)
[4] Aristotle. *Nicomachean Ethics. Book* II.1ff.
[5] Aristotle. *Nicomachean Ethics*. 1177b.
[6] Aristotle. *Nicomachean Ethics* X. vi–vii.7.

5.1 The Theoretic Life (βιος θεωρητικὸς)

held to excel in serious worth, consisting as it does in contemplation, and to aim at no end beyond itself, and also to contain a pleasure peculiar to itself...[7]

This last point concerning is important since the theoretic life will also partake in an eminent degree of the goods sought in the contending forms of life – the pleasure of the common man and the virtue of the political man. But the pleasures are not the physical pleasures favored and sought by the self-indulgent hedonist, but those of the very highest and most noble kind. These are the pleasures is of the highest and most noble faculty in man – the intellect.

As opposed to a political life – even one devoted entirely to moral virtue – the theoretic life is the most self-sufficient life possible for man on earth. For while the theoretic *man* will require the supply of necessities and hence the need for others, the activity of the theoretical life does not, and can be done even alone. By contrast the moral virtues proper to practical life *require* others for their very accomplishment – one cannot be generous, or just, or temperate without others whom one benefits.[8]

The idea that the transcendence of the polis and the need for the society of other human beings is a mark of the *superiority* of the theoretic life might seem quite ironic given how strongly in the *Politics* he emphasizes the idea that "man is by nature a political animal", and quotes Homer that the man without the city is "clanless, lawless, hearthless."[9] Yet we should also recall that Aristotle argues that the person who is naturally without a human community is *either* "low in the scale of humanity or above it."[10] Only a political life is properly *human*. Being apart from the Polis implies one might have descended to the beastly or ascended to the divine. This precisely the case Aristotle will make about the theoretic life:

> Such a life as this however will be higher than the human level: not in virtue of his humanity will he achieve it, but in virtue of something within him that is divine...[11]

The intellectual life is indeed for Aristotle the life proper to the gods and hence the intellect is that part of man which is most akin to the divine.[12] It is for Aristotle through the life of philosophy and its pure theoretic activity alone that man transcends the ties of the earthly community and ascends in a sense to a heavenly and divine form of life.

---

[7] Ibid. X.vii 5–7.
[8] Ibid. X.vii.4ff.
[9] Ibid. X.i.9.
[10] Idem.
[11] Ibid. X. vii.8.
[12] Ibid. X. viii.7.

5.2 The Substance of the Theoretical Life – The Theoretical Sciences

It remains to be seen what the content of the theoretic life is. In what kind of activity will it consist? Clearly, it will involve the theoretical sciences. We have already seen that the theoretical sciences in spite of their comparative inutility are deemed by Aristotle of a higher order than the practical sciences (which aim at action), or the productive sciences (which aim at the making of things). This is among other reasons due to the fact that they are concerned with the highest and most universal objects:

> ...the most honourable science must deal with the most honourable class of subject. The speculative sciences, then, are to be preferred to the other sciences...[13]

It remains then to enumerate the theoretical sciences, their subject, their hierarchy, and then also to examine the question of the highest and universal science which aims to unify and integrate all knowledge. First however it would be valuable to take a brief detour to consider one of Aristotle's signature achievements – logic – and its relation to the theoretic life.

5.3 The Role of Logic

Among Aristotle's contributions to intellectual history perhaps none was as evincing of his creative intelligence and far reaching influence as his invention of logic, considered as a formal science of validity. It remains then to briefly indicate what relationship if any exists between Aristotle's logic and his ideal of the theoretic life. Aristotelian is governed bounded by two contrasting tools of reasoning – induction and the deductive syllogism. *Induction* moves from knowledge of individuals to the universal. This form of reasoning has the advantage that it begins with what is more knowable to us insofar as sensation is the beginning of knowledge and sensation is of individuals and not universals (e.g. one *sees* individual dogs not "dog" as a universal concept).[14]

The Aristotelian discovery of inductive reasoning would furnish one of the most central foundations of the scientific revolution.; The empirical investigation of the scientist generally proceeds to formulate generalization by observation of particular cases – one for instance formulates a universal claim about what burning is and how it works by careful observations of many instances of burning processes. Because it begins by expansive and direct acquaintance with the individual things, Aristotle will acknowledge the principle point of Bacon – namely that induction is in the practical sense the more practically useful form of knowledge since practical action

---

[13] Aristotle. *Metaphysics*. E 1026a.
[14] Aristotle. *Metaphysics*. Δ 1018b32–34.

5.3 The Role of Logic

is directed to the individual and not the universal. Thus, for instance the person who knows from experience the individual patient may be better at curing that patient than a medical scholar who knows more about medicine in general. As Aristotle avers:

> It would seem that for practical purposes experience [knowledge of the particular] is in no way inferior to art [which pertains to the universal]; indeed we see men of experience succeeding more than those who have theory without experience. The reason of this is a that experience is knowledge of particulars, but art of universals; and actions and the effects produced are all concerned with the particular. For it is not man that the physician cures, except incidentally, but Callias or Socrates or some other person similarly named, who is incidentally a man as well. [20] So if a man has theory without experience, and knows the universal, but does not know the particular contained in it, he will often fail in his treatment; for it is the particular that must be treated.[15]

Remarkably -although in keeping with his anti-utilitarianism - however, Aristotle privileges *deduction* which moves from the universal to the particular. This is principally for three reasons.

First Aristotle is concerned above all with *demonstrative* knowledge. But "... demonstrations [ἀπόδειξις] are universal and universals cannot be perceived by the senses."[16] Sensation only yields knowledge of individuals which does not cover all cases of a thing – the fact for instance that all the birds one has seen can fly does not prove that all birds fly, merely that one has not yet seen the contrary case (e.g. penguins.). All contemporary theories in empirical science are parenthetically of this non-demonstrative form – even as well established a principle as "the maximum velocity that can be reached is the speed of light in a vacuum" would need to be revised if even a single contrary example were found. On the other hand, if one knows a universal truth – for instance that all squares have four sides, it follows by absolute deductive necessity that any individual square will have four sides. The greater part of Aristotle's logic then will pertain to the deductive syllogism.

Secondly, the greatest kind of knowledge to Aristotle is not that form which is most practically efficacious but that which provides knowledge of the universal cause. As he argues the laborer may know about the particular art of building from his practical experience, yet the architect is esteemed wiser on account of the fact that he understands the universal causes of the building as whole; likewise, the mother who knows from experience that chicken soup is good for her daughter when she is sick, is not by that fact as wise in medicine as the doctor who understands the universal principles of health. And as we have seen the highest science for Aristotle is that which deals with the most universal causes – metaphysics. In this way the hierarchy within Aristotle's logic conforms to his general privileging of the theoretic life. And yet Aristotle in his concern to establish the principles of a science of nature based on careful empirical observation, in his elaboration of the inductive method, and in his argument for its practical utility, Aristotle supplied crucial

---

[15] Aristotle. *Metaphysics*. A 981a. (My brackets)

[16] Aristotle. *Posterior Analytics*, 87b28 cf. https://www.loebclassics.com/view/aristotle-posterior_analytics/1960/pb_LCL391.157.xml (accessed May 10, 2018). (My brackets)

5.4 The Theoretical and Practical Sciences Distinguished

If the theoretic life will be that occupied with the theoretical sciences(θεωρετικαί), it remains to know what they are, and in particular what distinguishes them from the practical sciences. According to Aristotle the theoretical sciences moreover do not consider their object with a view to action but rather simply as truth for theoretic contemplation:

> The object of theoretic knowledge is truth, while that of practical knowledge is action...[17]

This does not mean that the practical sciences and the theoretical sciences may not at times intersect and study the same objects. For example, like the practical science of ethics it belongs also to theoretical science to study the good which includes"...the Good in each particular case and in general the highest Good in the whole of nature."[18] However, the two kinds of science even when they both consider the same object consider them in diverse ways, for the practical science considers it with a view to action:

> ...even when they are investigating *how* a thing is so, practical men study not the eternal principle but the relative and immediate application.[19]

The practical wisdom needed for concrete ethical judgements as we have seen depends not only on theoretical grasp of moral principles upon knowledge of particulars, which in turn depends upon experience. These kinds of variable contingencies however fall outside the realm of theoretical science which is concerned with those principles which are true universally and by necessity:

> Therefore in every case the first principles of things must be necessarily true above everything else –since they are not merely *sometimes* true...[20]

The highest and noblest sciences then will be those which are most concerned with the most universal and necessary causes and principles, and simply to understand their truth rather than with a view to their utility for action.

---

[17] Aristotle. *Metaphysics*. α 993 b.
[18] Aristotle *Metaphysics* A *982b*.
[19] Aristotle. *Metaphysics*. α 993 b.
[20] Idem.

5.5 The Division of Theoretical Science – Physics, Mathematics, and Theology

Having distinguished between the theoretical and practical sciences, it is also necessary to distinguish between the various theoretical sciences. Aristotle speaks of three which differ each according to their proper object.

Physics (φυσική) is for Aristotle the science which deals with nature (φύσις). The chief characteristic of all the kinds of things which fall Aristotle's "physics" studies is that of change and motion – "physics deals with mutable objects"[21] Hence though the genetic ancestor of modern concept of "physics", Aristotle's "physics" thus has a much broader meaning including the study of natural bodies. In short it is the whole domain of what we now call the natural sciences including but not limited to the most basic and general principles of change, matter, and motion.

Mathematics (μαθηματική) differs from physics for Aristotle in that "…some branches of mathematics deal with things which are immutable, but presumably not separable but present in matter."[22]) Aristotle is here implicitly distinguishing his position on the objects of mathematics from that of Pythagoreans and Platonists who treat mathematical objects as independently existing, immaterial realities. For Aristotle some characteristics as number, quantity, and shape inhere in natural objects, but merely can be *considered* by the mind as if they were independent objects. This distinction is elaborated in the *Physics* where Aristotle says of the mathematician that in treating mathematical objects:

> …he abstracts them from physical conditions ; for they are capable of being considered in the mind in separation from the motions of the bodies to which they pertain, and such abstraction does not affect the validity of the reasoning or lead to any false conclusions.[23]

The third theoretical science which Aristotle discusses is theology (θεολογική) which as we would expect is the science of the divine being or God. This is the science which treats of those realities which are both "both separable and immutable"[24] By this he means that the divine Being object which theology considers differs from both material and mathematical objects in being not only unchanging and eternal but also separate from matter and not only in intellectual consideration but in reality, Naturally for Aristotle this science will be the highest and supreme of the three he discusses in these passages since the divine being as above all change and materiality is the most noble. The fact that in a few short passages he calls this science the highest and primary one, and that God is the First Cause and Principle of all existence, occasioned much debate and commentary in the Middle Ages among both his Catholic disciples like St. Thomas Aquinas and his Islamic disciples like Avicenna

---

[21] Aristotle. *Metaphysics* E 1026a
[22] Idem.
[23] Aristotle. *Physics*. 193b.
[24] Aristotle. *Metaphysics* E 1026a.

(ibn Sina.) The question is whether theology is identical with the universal science.[25] This hermeneutic issue need not detain us, except to say that it would seem a universal science would seem to have to include the causes and principles of *all* the sciences within its purview, including of course the First Cause. With that it remains to consider the question of the universal science itself.

5.6 The Universal Science

At last we are left to consider "the supreme ambition of reason."[26] Aristotle's dream of a universal science oriented the Western intellectual tradition for centuries. At the dawn of modernity when Descartes, in spite of his breaks from Aristotelianism, strove not to overthrow this ideal but rather to set the universal science on a new foundation for the new era of the scientific revolution. The failure of this project in modernity and the consequent modern crisis of the sciences will be touched upon in future chapters. It remains now to consider Aristotle's own conception.

What does it mean for Aristotle to speak of a universal science? How does it differ from a "particular" science? Aristotle provides us with a pithy definition of both albeit pregnant with meaning to unpack:

> There is a science which studies Being *qua* Being [τὸ ὂν ᾗ ὄν], and the properties inherent in it in virtue of its own nature. This science is not the same as any of the so-called particular sciences, for none of the others contemplate Being generally *qua* Being; they divide off some portion of it and study the attribute of this portion...[27]

The "particular" sciences then are those which study a particular kind or "region" of being. Mathematics for example studies *quantitative* being. This conception can still readily be understood in relation to our contemporary taxonomy of the natural sciences. For example, biology studies *living* being. And we particularize sciences further through their subdivisions contained within the genus of living beings – zoology studies animal being, ornithology studies "bird-being", and so forth.

The universal science however, studies *being as such*. It includes therefore all reality within its purview. Such a science aims at the highest wisdom. As the knowledge of the universal and the causes pertains more to wisdom then knowledge of the particular and the effects, this science will involve the knowledge of the most universal causes. In short, the universal science will aim at *knowledge of the whole*.

This universal science will be also the *master science*. Through the study of universal "causes and principles" it provides the foundation for all the other sciences

---

[25] See for example Msgr. John F. Wipple. *Metaphysical Themes in Thomas Aquinas* (Washington D.C., CUA Press, 2007):39–44. Both Islam and Catholicism were also concerned to assert a science even higher than the "natural"theology of Aristotle, namely revealed theology.

[26] Cf. Jean Loup Seban "Philosophy" in *The Westminster Handbook to Reformed Theology*. (Louisville, KY: Westminster John Knox Press, 2001):169 Donald K. McKim ed. (Accessed May 5, 2018 via Amazon.com).

[27] *Metaphysics*. Γ 1003a. (My brackets)

and directs them. In will include both the knowledge of the good (which informs ethics) and the knowledge of things according to their essential natures. Hence it will aim at knowledge of the origin of things through their first causes and principles and at their ends. All the particular sciences are really branches of the master universal science, and thus in a sense their servants. Correspondingly this science of the most universal causes will be in effect the "ruler" of the other sciences:

> Inasmuch as Wisdom is the most sovereign and authoritative kind of knowledge, which the other sciences, like slaves, may not contradict, the knowledge of the *end* and of the *Good* resembles Wisdom…but inasmuch as it has been defined as the knowledge of the first principles and of the most knowable, the knowledge of the essence will resemble Wisdom.[28]

This is a highly ambitious ideal, for it is indeed the "function of the philosopher to be able to study all subjects"[29] and no branch of human inquiry or study can escape his purview. Through his engagement in the universal science, the theoretic life reaches its complex apex. In the mind man, the microcosm, the mind strives to contain the whole universe of within itself. Beholding that whole which it knows, it is in its very theoretic activity "…*becoming all things*…"[30] Though derided by the "practical man" as useless, the philosopher living the theoretic life most fully fulfills what it is to be human, i.e. the very function of man. For as Aristotle's great Catholic disciple would put the matter some 15 centuries later *propria autem operatio hominis inquantum homo est intelligere* – "The proper operation of man in so far as he is man is to understand".[31]

References

Aquinas, St. Thomas. *Sententia libri Metaphysicae*. Book I. Lectio I. At University of Navarra, Spain. http://www.corpusthomisticum.org/cmp0101.html. Accessed 4 May 2018.
Aristotle. *Nicomachean Ethics*. 1926. (1999 reprint). *Nicomachean Ethics*. Trans. H. Rackham, 1999 Loeb Classical Library, Harvard University Press.
———. 1933 (2003 reprint). *Metaphysics*. Trans. Hugh Tredennick. Loeb Classical Library, Harvard University Press.
———. *Posterior Analytics*. https://www.loebclassics.com/view/aristotle-posterior_analytics/1960/pb_LCL391.157.xml. Accessed 10 May 2018.
———. 1941. *Basic Works of Aristotle*. Trans. Richard McKeon. New York: Random House.
Seban, Jean-Loup. 2001. Philosophy. In *The Westminster Handbook to Reformed Theology*, ed. Donald K. McKim. Louisville: Westminster John Knox Press. Accessed 5 May 2018 via Amazon.com.
Wipple, John F. 2007. *Metaphysical Themes in Thomas Aquinas*. Washington, DC: CUA Press.

---

[28] Aristotle. Metaphysics. B 996b.

[29] *Metaphysics*. Γ 1004b.

[30] Aristotle. *On the Soul*. 430a14-14 (my ital.).

[31] St. Thomas Aquinas. *Sententia libri Metaphysicae*. Book I. Lectio I http://www.corpusthomisticum.org/cmp0101.html (accessed May 4, 2018).

Part II
The Baconian Revolt Against Greek Theoria and the Modern Birth of the Technological Mind

Fig. II.1 Sir Francis Bacon (1561–1626)
Sir. Francis Bacon from a 1738 engraving – contributed by Everett Historical. – www.shutterstock.com

> *Scientia et potentia humana in idem coincident...* (Bacon. *Novum Organum.* In Latin Library http://www.thelatinlibrary.com/bacon/bacon.liber1.shtml (Accessed June 1, 2018))
> Human knowledge and power coincide in the same thing...
> – Sir. Francis Bacon, *Novum Organum* I. III

> *Nostra profecto sunt antiqua tempora, cum mundus jam senuerit...* (For a Latin text of *De Augmentis Scientiarum.* see for example *The Works of Francis Bacon* . 1826. https://books.google.es/books?id=tTrjAAAAMAAJ (Accessed June 1, 2018))
> Our times are the ancient times, when the world is ancient... (Bacon. *The Major Works.*145.)
> – Sir. Francis Bacon. *De Augmentis Scientiarum. Book One*

Chapter 6
The Rebirth of Time: Sir. Francis Bacon and the Origins of Modernity

6.1 What Is Modernity?: The Birth of the Technological Mind

Any inquiry into the question "what is modernity?" must surely begin with something *distinctive* about modernity which breaks from the past. Many aspects of the pre-modern world from the study of classical Greek philosophy to the existence of Catholic monasticism have survived to the present day. But clearly none of these things are definitive of the modern era. Perhaps the most common answer to this question links modernity with the rise of *secularism*. The Islamic philosopher Seyyed Hossein Nasr provides us with an eloquent statement of this position:

> We, that is traditionalists like myself, use the term "modernism" not in a vague way as characterizing just things that happen to be around today, but as a particular way of looking at the world, a worldview that began in the Renaissance in the West…modernism rejects the primacy of absolute and ultimate truth transcending the human order and descending upon the human realm from the Divine Order. It places man himself at the center of the stage as the absolute.[1]

This answer has a certain *prima facie* plausibility. Religion has been a central feature of all the great world-civilizations in history, and indeed in some cases as with the Islamic civilization or the Hindu civilization, or for that matter the Catholic civilization of medieval Europe, religious tradition is the central and defining "sun" around which all other elements of the civilization rotate. Modern secularism therefore is something both radical and *distinctively modern*, which breaks from the whole standard pattern of the past.

As compelling as this answer is, secularism is rooted in something more primordial. Secularism is essentially *the belief in human self-sufficiency* and hence the lack of need or reliance on the divine. Man becomes as Nasr says the center of his own world. But this secular attitude did not appear until man first began to acquire *the*

---
[1] Seyyed Hossein Nasr. *In Search of the Sacred*, 182.

technological mind. Through technical mastery over his environment, and the ability to use technology to fulfill his desires, man made himself the center of his own universe. Secularism then appears from a historical perspective to be an *effect* of the technological project of mastery. Having once taking it upon himself to be the master of nature, modern man becomes intoxicated by his own power, the center of his own universe, and no longer feels the same need for God. If this is correct it would seem the originators of modernity, properly considered, are figures like Bacon and Descartes who though still inhabiting a religious universe prepared the path to the technological re-orientation of European culture.

Another view about the origins of modernity merits consideration is that of Leo Strauss. Strauss situates the rejection of the pre-modern tradition, and hence the inception of modernity in the political philosophy of Machiavelli. The classical tradition of political philosophy for Strauss is teleological, assuming man has an end of moral virtue or excellence for the sake of which politics exists. For Strauss the Machiavellian revolution was a radical break from this previous tradition. Strauss puts the matter

> Machiavelli rejects the whole philosophical and theological tradition…one must start from how men do live; one must lower one's sights. The immediate corollary is the reinterpretation of virtue: virtue must not be understood as that for the sake of which the commonwealth exists, but virtue exists exclusively for the sake of the commonwealth.[2]

Machiavelli is thus the figure who breaks with classical thought and inaugurates modernity by severing politics from classical ethics and the idea of natural right. Politics is concerned fundamentally with the acquisition of *power*, and what means are expedient to acquire and maintain it. For Strauss the new non-teleological science of Bacon comes in the wake of Machiavelli's revolution in political philosophy which inscribes the new science with its concern for power:

> The purpose of science is reinterpreted: *propter potentiam*, for the relief of man's estate, for the conquest of nature, for the maximum control, the systematic control of the natural conditions of human life.[3]

This view of the matter also is not lacking in evidence. Power was a central concern for the principal authors of the technological turn. This means for the true moderns in particular power over nature, though those who can master nature can also master men. Bacon famously said that *Scientia et potentia humana in idem coincident* "Power and knowledge coincide in the same thing",[4] while Descartes declared the goal of scientific knowledge for human beings to become *the masters and possessors of nature.*[5]

---

[2] Leo Strauss."The Three Waves of Modernity" in *An Introduction to Political Philosophy*, 86.
[3] Ibid, 88.
[4] Francis Bacon. *Novum Organum*, Aphorism I.3 http://www.thelatinlibrary.com/bacon/bacon.liber1.shtml (accessed 8/12/2018).
[5] Rene Descartes. *Discourse on Method*. 6 in *Discourse on Method and Meditations*. Elizabeth S. Haldane and G.R.T. Haldane(trans.) (Mineola, New York: Dover Publications, 2003):41 – accessed via Google Books. My italics.

6.1 What Is Modernity?: The Birth of the Technological Mind

Moreover Bacon clearly read and refers favorably refers to Machiavelli:

...hence we are beholden to Machiavel, and writers of that kind, who openly and unmasked declare what men do in fact, and not what they ought to do.[6]

Machiavelli's thought is thus crucially important to understand the modern turn. Yet, the argument for a Machiavelian origin to modernity overlooks the fact that Machiavelli's realism, his focus on power and expedience is really a revival on an older tradition rather than a pure innovation. This outlook or at least recognition of it, is to be found both in the Greek historians and in the Greek rhetorical tradition. Thucydides famously places into the mouth of the Athenian delegation to Melos the phrase "the strong do what they can, and the weak suffer what they must."[7] Those who favor the art of rhetoric meanwhile are presented by Plato as seeing its good to lie in the acquisition of power as a source of security if not as an end in itself. Callicles in the *Gorgias* appeals to nature itself for vindication of the idea that the way of things favors the strong and powerful, and not the just as ordinarily understood which is merely conventional:

...but nature in my opinion, herself proclaims the fact that it is right for the better to have advantage of the worse, and the abler of the feebler.[8]

Comparably, Thrasymachus proclaims in the Republic that "...the just is nothing else than the advantage of the stronger."[9] The "idealism" of the classical virtue tradition of the philosophers – of Socrates, Plato, Aristotle, the Stoics – which was bequeathed to the Roman and medieval worlds, was therefore itself *a reaction* to what Werner Jaeger calls "the philosophy of power."[10]

The key innovation of modernity then was to privilege *technology* in the service of power as the most important human activity and the chief means for *acquiring* power, in the Machiavellian sense, or otherwise. The dream of Bacon and Descartes was that the scientific knowledge of nature would render possible mastery over her through technology. Marx argued this dream of technological power over nature as fulfilled through the development of the capitalist economic order through the applied science of technology:

Subjection of Nature's forces to man, machinery, application of chemistry to industry and agriculture, steam- navigation, railways, electric telegraphs, clearing of whole continents for cultivation, canalizing of rivers...what earlier century had even a presentiment that such productive forces slumbered in the lap of social labour?[11]

From the standpoint of an ancient, the modern assertion concerning the primacy of technology would be profoundly irrational. Technology aims at the useful, and the useful, as we have seen with Aristotle, is only a subordinate and above all

---

[6] Bacon. *The Advancement of Learning* Book VII. http://oll.libertyfund.org/titles/bacon-the-advancement-of-learning (Accessed August 12, 2018).

[7] Thucydides, *The History of the Peloponnesian War*.III.89.

[8] Plato. *Gorgias*.483d (385 in Loeb).

[9] Plato. *The Republic*. 338d. (p. 47 in Loeb.)

[10] Werner Jaeger, *Paideia* II., 133.

[11] Karl Marx and Friedrich Engels. *The Communist Manifesto*, 245.

instrumental good. So if technology is the *primary* thing, it implies that there are no *ends*, but only *means*. But if everything is a means, and nothing is an end, then human activity as a whole is ultimately aimless. It also means the dethronement of the classical liberal arts which claim an inherent nobility beyond utility.

It is perhaps some such aimlessness and nihilism which Heidegger had in mind in his critiques of technology. Nonetheless, for Heidegger technology is more than merely instrumental. It is not *fundamentally* about tools – "…the essence of technology is by no means anything technological."[12]

It is the fundamental way in which beings are revealed to modern man:

> What is modern technology? It too is a revealing. Only when we allow our attention to rest on this fundamental characteristic does that which is new in modern technology show itself to us.[13]

How so? Because technology conditions for example the particular way in which nature shows itself to modern man. Specifically, nature reveals itself under the aspect of *resource*. Technology in its essence is about a *new relation to the earth*. "The earth now reveals itself as a coal mining district, the soil as a mineral deposit."[14] Technology is thus inextricably tied up with an *economic* perspective, reflected in the reduction of nature to economic evaluations.

Technology is also more than instrumental for Heidegger, because instrumentality implies subordination to human purpose, while for Heidegger technology has slipped beyond human purposes. A topic to which we shall return. Heidegger in many ways seems like the last in the line of German Romantics, most particularly Nietzsche, who wish to go back beyond technological rationality, and even metaphysical rationality, to a presumed original intuition of Being they find in the Greek poets and pre-Socratic philosophers.

At all events whatever one thinks of it, it appears that technology or rather the technological world view, is what now conditions everything about "modernity". And if this is modernity's inception, we must look first at Bacon in order to render modernity intelligible to us.

6.2 Bacon's Place in History

It is a curious matter of European intellectual history that the break from the middle ages commenced with a reclamation of classical antiquity and its ideals, but culminated in their repudiation. From this distinction we may take note of the chasm separating the world view of the Renaissance from "modernity" properly considered.

---

[12] Martin Heidegger. "The Question Concerning Technology" in Basic Writings, p. 311.
[13] Ibid., p. 320.
[14] Ibid.

6.2 Bacon's Place in History

In his 1533 work *De Pueris Recte Instituendis* the Renaissance cardinal and humanist Jacopo Sadoleto retold the famous story of the Greek scientist Archimedes who helped King Hiero defend his native Syracuse from a Roman siege using his expert knowledge of mathematics and physics to build machines of war. "And yet" Sadoleto avers:

> ...he was blamed for drawing into the area of common affairs, and so violently dishonouring an art [geometry] which owes its dignity mainly to its remoteness from the world of sense and sight..."[15]

Sadoleto seems to empathize with the critics who felt a liberal mind is one which:

> ...content with the sweetness of learning, seeks for nothing more, and is loath to suffer this serene delight to be interrupted by the rough clamour of popular applause.[16]

The Renaissance, whose ideals Sadoleto embodied, is often thought of as the beginning of modernity, and that is true in the sense that it represented a break with the medieval past. The boundless veneration of classical antiquity by the Renaissance humanists brought about an inevitable sense of distance from the "barbarous" middle ages. And yet Sadoleto's assumptions reveals the vast chasm that also separates the modern from the Renaissance mind. The belief that the application of rational knowledge to technological use is a kind of vulgarization is profoundly alien to the modern consciousness which tends to see its such technical utility precisely its primary value. The elevation of the pure philosophical contemplation of the highest and noblest things was native to the Greek world and integrated in a religious register into the medieval ideal of the *vita contemplativa*. Modernity begins at the moment when Aristotle's theoretic ideal is supplanted by a technological ideal – which is to say knowledge is placed at the service of practical utility. What explains this paradigm shift?

The overthrow of the empire of Aristotle and his ideal of the theoretic life over the European mind, rests to a remarkable degree on the work of a single historical figure – that of Sir. Francis Bacon.[17] Bacon himself was acutely conscious of living at the threshold of a new era "these times are the ancient times when the world is ancient."[18] For Bacon virtually all that could be gained from the study of the ancient philosophers like Plato and Aristotle had been already been extracted, and these worn paths had run their course. The veneration of antiquity characteristic of all previous phases of Western civilization and all civilizations –including the culture of the Renaissance – was for Bacon now holding back the advancement of

---

[15] Jacopo Sadoleto in *Sadoleto on Education. A Translation of the De Pueris Recte Instituendis – Primary Source Edition*. Ernest Trafford Campagnac (translator.) (Oxford University Press, 1916 – Nabu reprint): 124. (my brackets)

[16] Ibid. 125.

[17] For my exploration of relevant passages Bacon I am greatly indebted to Dr. James Nicholson, a neuroscientist with a deep interest in Bacon. Emphatically, this assistance in no way suggests his agreement with thoughts and conclusions.

[18] Francis Bacon. *The Advancement of Learning*. In Francis Bacon: The Major Works. (Oxford: OUP, 2008):145 – Herafter *The Major Works*.

knowledge. Bacon's notion of time itself is radical oriented for perhaps the first time in history not to the past but to the future – the inception of modern progressivism. No longer oriented to receiving the wisdom of the ancients, Bacon re-orients the Western mind for perhaps the first time in history toward the idea of future progress. To achieve this progress requires breaking free from the dead hand of tradition and the hold of authorities of the past:

> Again, men have been kept back as by a kind of enchantment from progress in the sciences by reverence for antiquity, by the authority of men accounted great in philosophy, and then by general consent.[19]

Central to Bacon's inauguration of modernity is his challenge to the authority of Aristotle. It is no coincidence that at the inception of modernity Francis Bacon styled his classic treatise the "new organon"(*Novum Organum*) setting it in deliberate contrast to Aristotle's *Organon*. Aristotle we must remember had acquired a unique status since the high middle ages. Long an authority in logic since the translations of Boethius, he became ever more central to the thought and culture of Catholic Christendom since the twelfth and thirteenth century when his works of natural philosophy, ethics, and metaphysics were translated and became authoritative (after some struggles) in the University of Paris. It is true that Catholic scholasticism in the great work of synthesizing the Aristotelian corpus with Catholic theology had in a sense subordinated philosophy to the role of *ancillia theologiae* ("the handmaiden of theology"), thus making the "master" science of Aristotle a servant of the yet higher science of theology. And to be sure the purely natural end of man envisioned by Aristotle stood in a certain situation of tension with the Christian idea of a supernatural end beyond this world. Roman Catholicism could not accept the idea of a purely natural end of man, attainable through natural powers, as this would seem to render the order of grace and redemption superfluous. Boethius of Dacia in his *De Sumo Bono*[20] who seemed to flirt with an Aristotelian naturalism in a Christian context by suggesting a beatitude attainable by natural exercise of the intellect accessible to the philosopher. This was part of the issue which resulted in the famous 1277 Paris condemnations. Yet, St. Thomas Aquinas will go quite far as to interpret Christian salvation on terms deeply influenced by Aristotelian intellectualism, seeing the vision of the divine essence – the *visio beatifica* as the supernatural fulfillment of man's natural desire to know.[21] And with-

---

[19] Francis Bacon. *Novum Organum*. Aphorisms Book One, LXXXIV. https://constitution.org/bacon/nov_org.txt (Accessed August 13, 2018).

[20] This work can be found here. http://www.documentacatholicaomnia.eu/03d/1200-1300,_Boetius_Dacius,_De_Summo_Bono_Sive_de_Vita_Philosophi,_LT.pdf (Accessed 4/28/18). On the 1277 condemnations see for instance F.C. Copleston. *A History of Philosophy*. Volume II. (New York: Image-Doubleday, 1993):441.

There is a much larger extensive bibliography on this topic which falls outside the scope of the present treatment.

[21] See St. Thomas Aquinas. *Summa Theologiae* I-II, Q. 3, Article 8. The problem of a natural desire

For a supernatural end has raised issues of great complexity within Catholic theology, which fall beyond the scope of this present work.

6.2 Bacon's Place in History

out doubt from the thirteenth century until the scientific revolution, Aristotle was considered the supreme authority in natural knowledge with a singular empire over the European mind. He was *hoc philosophus* (*the* philosopher) to St. Thomas Aquinas, and "the master of those who know" to Dante. While it required modification to assimilate Aristotle's ideal of the theoretic life to the ideal of the *vita contemplativa* embodied in the Catholic monastic life, the supremacy of the contemplative over the active life was not itself in question.

Given this close association of Aristotelianism with Roman Catholic scholasticism, the Protestant Reformation as one might expect created the possibility for a sense of discontinuity from Aristotle's authority. Martin Luther himself took aim at the scholastic synthesis of Christian theology and Aristotelian philosophy complaining that:

> As though we had not the Holy Scriptures, in which we are abundantly instructed about all things, and of them Aristotle had not the faintest inkling! And yet this dead heathen has conquered and obstructed and almost suppressed the books of the living God, so that when I think of this miserable business I can believe nothing else than that the evil spirit has introduced the study of Aristotle.[22]

While medieval Catholicism had accorded the system of classical philosophy represented by Aristotle a unique authority, Protestantism legitimated a broader sense of discontinuity from the medieval tradition of philosophy. With its emphasis on "faith alone" and a rather grim view of human powers after the fall, the Catholic synthesis of faith and philosophical reason was re-appraised. Yet in natural philosophy nothing had yet replaced Aristotelianism which remained in direct and indirect ways remained a central intellectual influence in Protestant England. Formidable examples include the work of John Case whose *Lapis Philosophicus* (1599) emphasized the authority of Aristotle, or the supreme Anglican theologian of the Elizabethan era, Richard Hooker who was deeply influenced by the medieval Thomistic tradition.

For Bacon this "servile" reliance on the authority of figures like Aristotle was stifling to the progress of the sciences. In his *Redargutio Philosophiarum* he writes:

> But even though Aristotle is the man he is thought to be I should warn you against receiving as oracles the thoughts of one man. What justification can there be for this self-imposed servitude?

And shortly later:

> ...assert yourselves before it is too late. Apply yourselves to the study of things themselves. Be not forever the property of one man.[23]

---

[22] Martin Luther. *An Open Letter to the Christian Nobility*. http://www.iclnet.org/pub/resources/text/wittenberg/luther/web/nblty-07.html (accessed Nov 5, 2015).

[23] *Redargutio Philophiarum*. Sp. III, 568–9 –
quoted in Rossi. P. 60. Can be found also in https://books.google.es/books?id=L-mOtwioWYAC&pg=PA49&dq=%22...assert+yourselves+before+it+is+too+late,%22+Bacon&hl=en&sa=X&ved=0ahUKEwid077pzPvaAhXD7RQKHZMQCTcQ6AEILzAB#v=onepage&q=%22...assert%20yourselves%20before%20it%20is%20too%20late.%22%20Bacon&f=false (accessed May 10, 2018).. Excellent also is Rossi's article on Baconianism in *The Dictionary of*

The truly revolutionary significance of Bacon does not rest principally as is often thought on his contribution to the development of a scientific method. As impressive an instrument as the new methods are, they represent no sharp break from ancient and medieval traditions of natural philosophy. We have already seen *Aristotle's* importance in the development of the induction and his argument for its superior utility. Platonism which informed the Patristic theology of the Roman Catholic Church emphasized the knowledge of the eternal and immutable Forms, and saw the natural order as one of mere "opinion"(δοχα) lacking stable being. Aristotle however restored interest in the natural world of change and in the observation and categorization of natural phenomenon by empirical methods.

Logically the Aristotelian revolution of the twelfth and thirteenth centuries restored interest in the natural world of change and flux. If Aristotelianism to the medieval scholastics served principally as an *ancillia theologiae*, nonetheless it also established the place of natural philosophy and a new concern for methods of observation including Aristotle's own inductive approach. Roger Bacon for example most famously devoted the sixth part of his *Opus Maius* (1267) to experimental science.

The emphasis on the empirical, singular realities received even great emphasis in the late fourteenth century with the work of William of Ockham credited with the rise of nominalism. Platonic and Aristotelian philosophy had emphasized universal concepts. Ockham argued that universal concepts while useful were really linguistic constructions of the mind – *Quod enim nullum universale sit aliqua substantia extra animam exsistens evidenter probari potest.*[24] "That there is no universal which is a substance existing outside the mind can be evidently proved."

The fundamental realities are the singular existents perceived by the senses, which are therefore the true objects of knowledge. Nominalism which was particularly strong in Britain, tended naturally to shift the emphasis in obtaining knowledge to induction which as we have seen begins with singulars. Francis Bacon's fruitful concern with the inductive method can thus be read as an organic development from the medieval scientific tradition itself ultimately rooted in Aristotelianism itself. Bacon himself, the great prophet of modern science and ostensible enemy of Aristotle acknowledges his debt to Aristotle in a letter to Lord Mountjoye which prefaces his essay *Of The Colours of Good and Evil:*

> …I am happy to do the part of a good house-hen, which, without any strangeness will sit upon pheasants' eggs. And yet, perchance, some that shall compare my lines with Aristotle's lines, will must by what art, or rather by what revelation, I could draw these conceits out of that place. But I, that should know best, do freely acknowledge, that I had my light from him ; for where he gave me no matter to perfect, at the least he gave me occasion to invent.[25]

Bacon's true radicalism is not primarily then a question of *method*.

---

the History of Ideas. http://xtf.lib.virginia.edu/xtf/view?docId=DicHist/uvaBook/tei/DicHist1.xml;chunk.id=dv1-25;toc.depth=1;toc.id=dv1-25;brand=default (accessed 5/30/2016).

[24] William of Ockham. *Summa Logicae.* 14:16 http://www.logicmuseum.com/wiki/Authors/Ockham/Summa_Logicae/Book_I/Chapter_15 (Accessed November 2, 2015).

[25] Francis Bacon. *Of The Colours of Good and Evil:* In https://archive.org/stream/worksfrancisbaco02bacoiala/worksfrancisbaco02bacoiala_djvu.txt - (accessed date 11/12/2015). Recently, http://clarityonthesea.org/files/pdf_archive/bacon-1824-works%20of%20francis%20bacon_2.pdf (10/27/2018)

To understand the nature of Bacon's revolution it might be useful to recall Aristotle's defense of the theoretic life as the good life. Aristotle's thesis rests as we have seen on three principal conceptual pillars.

First the idea wisdom is a supreme good in itself possessing its own proper excellence apart from any utilitarian consideration. From this it follows logically that the pursuit of wisdom (i.e. philosophy) will be the best and noblest form of human life.

Secondly, from this principle follows a hierarchy of the sciences. If the quest for wisdom is in and of itself is the supreme human pursuit, that science which seeks the highest and most universal wisdom – first philosophy (i.e. metaphysics) – will be the supreme science Its lack of utility proves rather than detracts from its value, for the highest science like the highest good serves nothing beyond itself. The lowest sciences will be those which serve utilitarian aims such as the so called mechanical sciences.

Thirdly, Aristotle defends the supremacy of the theoretic life on the basis of his natural teleology. Reason constitutes the "function" or purpose of man, and consequently the good life will be the life of reason. The life of philosophical contemplation even when remote from practical affairs is the purest expression of the life of reason. Consequently it is the theoretic life which most perfectly fulfills the end of man.

Each of these contentions comes under radical scrutiny by Bacon and it is this critique which we must understand. Because the third principle of teleology involves matters of great complexity in cosmology and ethics, and because his critique of teleology was amplified by other early modern thinkers such as Descartes and Spinoza, we shall focus in this chapter on the first and second issues – Aristotle's non-utilitarian view of knowledge as well his hierarchy of the sciences. Let us take up each of these issues in turn.

References

Aquinas, St. Thomas. 1947. *Summa Theologica*. Trans. Dominican Fathers of the English Province. At DHS priory. https://dhspriory.org/thomas/summa/FS/FS003.html#FSQ3OUTP1. Accessed May 2018.

Bacon, Francis. 2008. *Francis Bacon: The Major Works*. Oxford: Oxford University Press.

———. *De Augmentis Scientiarum*. 1826. In *The works of Francis Bacon*. https://books.google.es/books?id=tTrjAAAAMAAJ. Accessed 1 June 2018.

———. *Novum Organum*. 1620 (English trans. Based on Robert Leslie Ellis 1863.) http://www.thelatinlibrary.com/bacon/bacon.liber1.shtml. Accessed 1 June 2018.

———. English trans. of *Novum Organum* (Based on Robert Leslie Ellis 1863.) http://www.constitution.org/bacon/nov_org.htm. Accessed August 2018.

Boethius of Dacia. *De Summo Bono*. In Documentacatholica.eu http://www.documentacatholicaomnia.eu/03d/1200-1300,_Boetius_Dacius,_De_Summo_Bono_Sive_de_Vita_Philosophi,_LT.pdf. Accessed May 2018.

Copleston, F.C. 1993 (reprint). *A History of Philosophy*, Vol. II. New York: Image Doubleday.

Heidegger, Martin. 2008. *Basic Writings*, ed. David Farrell Krell. London (et al.): HarperCollins.

Luther, Martin. 1915. *The Works of Martin Luther*. Philadelphia: A.J. Holman Company Internet Christian Library. http://www.iclnet.org/pub/resources/text/wittenberg/luther/web/nblty-07.html. Accessed May 2018.

Marx, Karl and Engels, Friedrich. 2002 (1848). *The Communist Manifesto*. Gareth Stedman Jones (ed.), Samuel Moore (trans). London (et al.): Penguin.

Nasr, Seyyed Hossein. 2010. *In Search of the Sacred*. Santa Barbara, et al: Praeger. Interview with Ramin Jahanbegloo. Accessed 13 Aug 2018. via JHU library.)

Rossi, Paolo. 1968. *Francis Bacon from Magic to Science*. Trans. Sascha Rabinovitch. London: Routledge & Kegan Paul.

———. 1973. Baconianism. In *Dictionary of the history of ideas*. Vol. I. 172–179. Charles Scribner's Sons. At University of Virginia Library. http://xtf.lib.virginia.edu/xtf/view?docId=DicHist/uva-Book/tei/DicHist1.xml;chunk.id=dv1-25;toc.depth=1;toc.id=dv1-25;brand=default. Accessed May 2018.

Sadoleto, Jacopo. 1916 translation. (Nabu reprint.) *Sadoleto on Education. A Translation of the De Pueris Recte Instituendis* Trans. Ernest Trafford Campagnac (translator.) Oxford University Press, 1916 – Nabu reprint.

Sheehan, Thomas. 2013. *Thomas Sheehan on Heidegger and Technology*. Podcast originally at Stanford: http://french-italian.stanford.edu/opinions/ (I believe I accessed a You Tube version in August, 2018).

Strauss, Leo. 1989. The Three Waves of Modernity. In *An Introduction to Political Philosophy: Ten Essays by Leo Strauss*, ed. Hilail Gilden. Detroit: Wayne State University Press.

William of Occam. *Summa Logicae*. Logic Museum. http://www.logicmuseum.com/wiki/Authors/Ockham/Summa_Logicae/Book_I/Chapter_15. November 2, 2015.

Chapter 7
Bacon's New Magic: The Transfigured Aim of the Sciences

As we have seen Aristotle's view of the sciences is characterized by an anti-utilitarian view which regards the quest for wisdom as an end in itself. Hence his claim that metaphysics is at once the most excellent and least useful of all sciences. The radical quality of Bacon's thought begins with his effort to entirely re-orient the aim of the sciences from what the classics understood. Bacon puts it plainly that the aim of knowledge ought to be concrete improvement of the human condition and not the mere satisfaction of wonder.

> But the greatest error of all the rest is the mistaking or misplacing of the last or furthest end of knowledge. For men have entered into a desire for learning sometimes upon a natural curiosity or inquisitive appetite…as if there were sought in knowledge a couch, whereupon to rest a searching and restless spirit…and not a rich storehouse, for the glory of the Creator and the relief of man's estate.[1]

In order to improve the estate of man is it necessary to master nature and unlock its secrets. In his *Novum Organum* ("instrument") – consciously titled as an antitype to Aristotle's Organon – he famously writes:

> Human knowledge and human power meet in one; for where the cause is not known the effect cannot be produced. Nature to be commanded must be obeyed; and that which in contemplation is as the cause is in operation as the rule.[2]

For Bacon then is the aim is augmentation of *human power*. And by power, Bacon means primarily power over nature, or in short technology. The goal of establishing man's dominion over nature rests on the advancement of the sciences for only by understanding nature can one conquer it:

> But if a man endeavor to establish and extend the power and dominion of the human race itself over the universe, his ambition (if ambition it can be called) is without doubt a more wholesome and a more noble thing than the other two. Now the empire of man over

---

[1] Francis Bacon. *The Advancement of Learning*. In *The Major Works*, 147–148.
[2] Bacon. 1960. *The New Organon*. Aphorisms I.iii, 39.

things depends wholly on the arts and sciences. For we cannot command nature except by obeying her."³

Bacon will use quite aggressive language in describing the human relationship to nature. The secrets of nature [*occulta naturae*] must be as it were violently forced out of her if she is to be conquered and mastered for- "…the secrets of nature reveal themselves more readily under the vexations of art [*vexationes artium*] then when they go their own way."⁴

The quest for knowledge is therefore subordinated to a practical goal -"the relief of man's estate."⁵ Bacon is thus an archetypal "prophet" of the modern technological civilization which aims at improving the human condition through the domestication of natural forces⁶ and harnessing them to human advantage.

Baconian science has little to do with the quest for knowledge which animated the Greek philosophers. We may recall that Xenophon's Socrates explicitly considers – and rejects – the idea of power over nature as an end of science:

> Do those who pry into heavenly phenomena imagine that, once they have discovered the laws by which these are produced, they will create at their will winds, waters, seasons, and such things to their need? Or have they no such expectation, and are they satisfied with knowing the causes of these various phenomena?⁷

Awe have seen with Bacon's inductive *methodology* it is in essence an organic development from the experimental concerns of Aristotelianism and medieval natural philosophy; yet his *aim* has more relationship to the tradition of *magic* which seeks knowledge of hidden things for the sake of wondrous and commanding power. As he writes:

> We here understand magic in its ancient and honorable sense – among the Persians it stood for a sublimer wisdom, or a knowledge of the relations of universal nature, as may be observed in the title of those kings who came from the East to adore Christ. And in the same sense we would have it signify that science, which leads to the knowledge of hidden forms, for producing great effects…⁸

---

³Ibid. CXXIX, p.118–119

⁴Ibid. Aphorisms Book I XCVIII(98), p. 95. (My brackets). The Latin phrase is et occulta naturae magis se produnt per vexationes artium, quam cum cursu suo meant. http://www.thelatinlibrary.com/bacon/bacon.liber1.shtml (accessed May 2018). There is an ample secondary literature concerning debates as to how best to understand "vexationes" which is sometimes understood as torments or tortures cf. "Carolyn Merchant. "Francis Bacon and the 'Vexations of Art': in *The British Journal for the History of Science* 46, no. 4 (2013): 551–99.

⁵Bacon. *The Major Works*. 147.

⁶I am indebted to my former teacher Jean Loup Seban for the concept of the "domestication of nature" which likely has other sources.

⁷Xenophon. *Memorabilia*. 1.15 (E.C. Merchant Translation) in http://www.perseus.tufts.edu/hopper/text?doc=Perseus%3Atext%3A1999.01.0208%3Abook%3D1%3Achapter%3D1%3Asection%3D15

Cf. also Leo Struass. *On Tyranny* (Chicago, Illinois: University of Chicago Press, 2000):178.

⁸The Advancement of Learning. III.5 – http://oll.libertyfund.org/titles/1433 (accessed Dec. 6, 2015).

In fact Bacon's ambivalent relationship to magic has been an object of interest to scholars who have explored his connections to theoreticians of the European magical tradition.[9] One of its foremost representatives Cornelius Agrippa in his work *De Occulta Philosophia Libri Tres* (1533) speaks of man as a magician *magus* and magic as the knowledge of secret things which will make him the dominator of the natural order.[10] This kind of magic as Agrippa says "…applies the hidden forms to the production of wonderful operations."[11] Agrippa was certainly read by Bacon.[12] Of particular importance for Bacon is the tradition of natural magic represented by Giambattiista Della Porta's text *Magia Naturalis* (1558) which emphasized magic drawing out of nature its secret powers rather than magic than through concourse with supernatural powers. Here the idea of magic converges with the broader tradition of natural philosophy.[13]

Now it is true Bacon was also a *critic* of the magical tradition, but his critique was based principally on its *fruitlessness*:

> Again the students of natural magic, who explain everything by sympathies and antipathies, have in their idle and most slothful conjectures ascribed to substances wonderful virtues and operations; and if ever they have produced works, they have been such as aim rather at admiration and novelty than at utility and fruit.[14]

Thus where Bacon turns his attention to alchemy he is sympathetic to the goal even if he regards its methods as at best haphazard. Yet making discoveries randomly and accidentally while looking for other things rather than augmenting knowledge systematically. He says of it that "…the search and stir to make gold hath brought to light a great number of good and fruitful inventions as well for the disclosing of nature as for the use of man's life."[15]

Could the same value for "the use of man's life" in Bacon's view be said of the study of the ancient philosophers? And since utility is what is important for Bacon one could argue he would see the blundering tinkering of the alchemist as of more value than the study of Aristotelian metaphysics. Bacon's project therefore is *to conjoin the aim of magic* – that of augmenting human powers and dominating nature – *with a systematic, scientific method* for attaining it.

Bacon's conception of the aim of science as the augmentation of human powers has a nexus also in the Christian humanism of the Renaissance with its exalted view of human possibilities and its emphasis on human dominion over nature. Consider for instance Tomasso Campanella's vision which proclaims man:

> A second God, the First's own miracle, he commands the depths; he mounts to Heaven without wings and counts its motions and measures and its natures. The wind and the sea he

---

[9] Cf. Rossi. *From Magic to Science*.
[10] See. Francis Yates. *Bacon's Magic* http://www.nybooks.com/articles/1968/02/29/bacons-magic/ (accessed 12/7/2015) – a review of Rossi's book.
[11] Quoted in Rossi, 21.
[12] Cf. Bacon. *A Letter to Sir. Henry Suville* in *The Major Works*, 118.
[13] Cf. Rossi. 19.
[14] Bacon. *The New Organon*.(New York:Macmillan, 1960): p. 83 – Aphorisms I. LXXXV.
[15] Bacon. *The Advancement of Learning*. Book One. *The Major Works*, p. 143.

has mastered and the earthly globe with pooped ship he encircles, conquers, and beholds, barters and makes his prey.[16]

The religious aspects in the thought of both Campanella and Bacon touches on an important point. One often associates modernity particularly with the issue of secularization. Yet, is notable that the first pioneers of the modern technological project such as Bacon and Descartes were themselves pious Christians.

Here we have the significance of *the technological ideal* as the true magic which has entranced and dominated Western modernity. Who can today deny that man has achieved the conquest of nature and laid bare its secrets, learned to fly like the birds, commands the elements, and communicate at the speed of light?

However, the subordination of knowledge to the aims of a technological project involves an implicit critique of the ideal of the theoretic life which sees knowledge itself as its own end. The radicalism of Bacon's project here bears mention.

It is instructive in this regard to consider Bacon's critique of the ancient philosophical system as *fruitless*. This is not a critique which even makes sense on the terms of Aristotelianism, where as we have seen utility is an indication of subordinate value. In his *Redargutio Philosophiarum* Bacon by contrast writes:

> There is no 'sign' more certain and more noble than that from fruits. In religion we are warned that faith is shown by works. It is altogether right to apply the same test to philosophy. If it be barren let it be set at naught. All the more should this be so if instead of the fruits of grape or olive, it bear the thistles and thorns of disputes and contentions.[17]

What we see here is the return of *utility* as the normative standard for evaluating knowledge. The radicality of this position as it bears on the problem of the theoretic life bears mention. For Aristotle the intellect or reason is the highest element of the human soul, and that which distinguishes man from other animals. It follows from this proposition that the form of life which most cultivates and perfects man's rational nature – that is the theoretic life – is the best and noblest form of human life. Wisdom even apart from any utility or effects is the supreme good.

If Bacon is correct that knowledge is best judged by its "fruits" – i.e. its useful effects this involves a radical reorientation away from the Aristotelian philosophic ideal. Knowledge in this case subordinated to other ends such as "the relief of man's estate" or the augmentation of human power. In that case material improvements of the human condition or the practical increase of human power over nature over what Aristotle finds in the perfection of the soul adorned with wisdom and virtue.

Bacon's critique of Aristotle and other ancients does not rest principally on an elaborate intellectual refutation of their philosophical positions, but rather on his *premise* concerning the importance of science bearing useful fruits. The systems of Greek metaphysics stand condemned most of all for their fruitlessness for "the relief of man's estate" or the augmentation of his powers. Neither the great, ancient systems of Plato and Aristotle, nor the disputations of the scholastics have practi-

---

[16] Quoted in Christopher Dawson. *Christianity and European Culture. Selections from the Work of Christopher Dawson.* Gerald Russello (ed.). (Washington, D.C.: C.U.A. Press, 1998): 176.
[17] As quoted in Rossi. *Bacon: From Magic to Science*, 49.

cally speaking improved the human condition or enhanced man's control of nature. So what, asks Bacon, is their value?

Yet this whole Baconian critique considered by itself rests on a *petitio principii*. Perhaps the principles of Aristotelian metaphysics can be refuted as false on their own terms as contradictory or contrary to the true nature of reality. But the mere fact that Aristotle's metaphysical claims are not useful in a pragmatic-technologically sense has no probative value in refuting them unless one first accepts the premise that truth is reducible to utility or at least all knowledge ought to be useful, and consequently that the absence of use-value argues for their falsity.. This requires that one sideline without confronting seriously all the arguments adduced by Aristotle on behalf of the intrinsic excellence of knowledge and the subordinate value of utility as a claimant to "the good." It is on this issue that the claims about the theoretic life turns; if Aristotle is correct than a life devoted to the philosophical contemplation the highest truths even apart from any practical result is the noblest life of which man is capable. If Bacon is correct than the value of knowledge is to be found by its fruits – does it affect practical improvements of the human condition? Each of these views will lead to a differing conception of the hierarchy of the sciences. For Aristotle the highest science will be First Philosophy (i.e. metaphysics) which seeks the most universal wisdom though being good in itself it is also the least useful; the lowest sciences will be the mechanical arts which have no real good in themselves but are good only for their utility in producing something belong themselves. One would engage in philosophy (but not for instance, agriculture or commerce) for its own sake. Bacon's privileging of utilitarian goals in knowledge will necessarily involve a radical re-evaluation of the hierarchy of the sciences and of the mechanical arts (technology), in particular.

References

Bacon, Francis. 1620 (1960). *The New Organon*, ed. Fulton H. Anderson. New York: Macmillan Publishing Company.
———. *At the Latin* Library. http://www.thelatinlibrary.com/bacon/bacon.liber1.shtml. Accessed 14 July 2012.
———. 2008. Francis Bacon: *The Major Works*. Oxford: Oxford University Press.
———. *At the Liberty Fund*. http://oll.libertyfund.org/titles/1433. Accessed 6 Dec 2015.
Dawson, Christopher. 1998. *Christianity and European Culture*. Selections from the Work of Christopher Dawson, ed. Gerald Russello Washington, DC: CUA Press.
Merchant, Carolyn. 2013. Francis Bacon and the 'Vexations of Art': Experimentation as Intervention. *The British Journal for the History of Science* 46 (4): 551–599. https://doi.org/10.1017/S0007087411000665.
Xenophon. 1923 (2013 revision). *Memorabilia. Oeconomicus. Symposium. Apology*. Trans. E.C. Marchant. O.J. Todd. Loeb Classical Library, Harvard University Press.
Yates, Francis. 1968. *Bacon's Magic*. In New York Review of Books. http://www.nybooks.com/articles/1968/02/29/bacons-magic/. Accessed May 2018.

Chapter 8
Technology Displaces Metaphysics – Bacon's New Hierarchy of the Arts and Sciences

As we have seen Aristotle's hierarchy of the sciences rests upon his distinction of goods. Goods which exist for the sake of some other good are inferior to goods which exist for their own sake. Consequently, goods of utility are inferior to those noble in themselves. From this it follows that the highest sciences will not the most useful but those whose object possesses the highest intrinsic nobility. At the apex then of Aristotle's hierarchy of the sciences then is metaphysics which seeks the highest and most universal wisdom. It is supreme in excellence not in spite but *because* it is the least useful for like the free man it exists for its sake and not for the sake of another. Metaphysics in extending reason to embrace all Being and Reality is the perfection and completion of man's rational nature. In general terms the theoretical sciences (which include also mathematics and natural science as Aristotle makes clear Book Epsilon of his *Metaphysics*) are superior to the practical sciences, as is the theoretic life to the practical-political life.

Corresponding to these evaluations are Aristotle's depreciation of the "banausic" mechanical activities such as manual labor, trade, and commerce. These things do not pertain to man's essential nature, nor are they done for their own sake but "merely" as useful to some other end. These activities if allowed to dominate the whole of human life obstruct the possibility of fulfilling man's nature as an intellectual being and may even for Aristotle render him unfit for civic participation. And in spite of their more practical culture, in Roman times the gentleman continued to regard manual labor, trade, and commerce as unworthy of those of high station as we have seen for instance in Seneca for whom the liberal arts alone possess the intrinsic excellence worthy of free men.

Christianity introduced some dignity into the status of the mechanical arts by giving manual labor greater value – Christ after all was a carpenter. For Hugh, even the mechanical arts come under philosophy in the broadest sense.[1] We may also consider for instance the Benedictine motto *ora et labora* – pray and labor. In

---

[1] Op. Cit. 752 – Hugh of St. Victor says ...*philosophia aliquo modo ad omnes res pertinere dicitur*– "philosophy in a certain way is said to pertains to all things". (my trans.)

another sense however the Aristotelian hierarchy of contemplative over practical activity in Aristotle harmonized well with the medieval ideal of religious life embodied in the monastic and mendicant orders. Though Aristotle's philosophical-theoretic life is not precisely the same as the Catholic religious-contemplative life, it is notable that St. Thomas Aquinas references the authority of Aristotle in making his defense of the supremacy of the *vita contemplativa* over the *vita activa*.[2]

It is again to Bacon that we must turn for a genuine revolution in thought. From Bacon's re-conception of the end of knowledge there follows a re-conception of the hierarchy of the sciences. For if knowledge does not have an *intrinsic* value as the exercise and perfection of man's rational nature, if instead it is justified by its utility for augmenting human power and improvement of human life, naturally the mechanical arts must be conceived in an entirely different light:

> But if my judgement be of any weight, the Use of History Mechanical is of all others is the most radical and fundamental towards natural philosophy; such natural philosophy as shall not vanish in the fume of subtile, sublime, and delectable speculation, but such as shall be operative to the endowment and benefit of man's life.[3]

So far from accepting Aristotle's hierarchy whereby the speculative sciences are distinct and superior to the practical and mechanic ones, Bacon indeed with take aim at the whole distinction declaring that the speculative and practical work in concert:

> Neither the naked hand, nor the understanding left to itself can effect much…And as the instruments of the hand either give motion or guide it, so the instruments of the mind supply suggestions for the understanding or cautions.[4]

Bacon's famous maxim that human knowledge and human power are essentially the same (or "coincide" in the same), is testifies also to his insistence on the unity of the speculative and the operative aspects of the sciences. This insight is of course essential to the modern development of scientific method which combines speculative hypothesis with practical observation (often technologically augmented). It presumes of course that the goal is to reproduce the effect.

Corresponding then to Bacon's relative exaltation of the mechanical sciences (i.e. technology) is his relative demotion of the highest science in Aristotle's hierarchy – metaphysics. Thus, Bacon contrasts the progressive character of the history of mechanical science against the static if not degenerating character of philosophy which according to Bacon have made virtually no progress since ancient times:

> So we see, artillery, sailing, printing and the like, were grossly managed at first, and by time accommodated and refined; but contrariwise the philosophies of Aristotle, Plato,

---

[2] Cf. St. Thomas Aquinas. *Summa Theologica*. ST. II-II, Q. 182.
[3] From the Advancement of Learning (Book Two) In Francis Bacon: The Major Works. 178.
[4] "Nec manus nuda, nec intellectus sibi permissus, multum valet…Atque ut instrumenta manus motum aut cient aut regnum; ita et instrumenta mentis intellectui aut suggerunt aut cavent." Latin text can be found at http://www.thelatinlibrary.com/bacon/bacon.liber1.shtml (accessed January 16, 2016). English text is from Bacon's *New Organon*.(F. Anderson ed.),39.

Democritus, Hippocrates, Euclides, Archimides, of most vigour at the first, and by time degenerate and imbased...[5]

Here early modern Europe's material and technological advances are crucial context for Bacon's perspective.[6]

A second and perhaps more substantive critique of Aristotelian metaphysics in Bacon is his critique of the doctrine of formal and (most significant for us) final causality which underpins it. In this as we shall see he is joined with other leading early modern thinkers like Descartes and Spinoza. This is a matter of significant importance for our consideration of the theoretic life, because it is Aristotle's contention that the theoretic life of philosophical contemplation constitutions the end of man and the perfection of his (rational) nature. If natural teleology is overthrown however than nothing in nature has a "natural end" and consequently man himself has no natural end. Teleology is the great bond which links Aristotle's physics and metaphysics to his ethics and politics. The moderns aim to sever this bond. Were they successful? This is the topic of our next chapter.

Bacon's revolution has of course borne fruit beyond anything he himself could have imagined. The technological revolution the Baconian project envisioned has dramatically transformed the way the average human life is lived. And yet from the perspective of philosophy it remains undergirded more by sheer assertion than by demonstration. Baconian philosophy helped to stimulate the advances in communications, transport, production, medicine, etc.... with which we are familiar. The relative neglect of the mechanical arts needed for such convenience is this to a degree a valid critique of the ancients and medieval. But are these ends in themselves? What can it mean to say that the end of knowledge is nothing beyond the augmentation of human power or the practical improvement of man's material condition? The increase of technological prowess involves an increase in *both* man's creative and destructive power (as became evident in the twentieth century.) Power therefore cannot have the character of an *intrinsic* good.

And what could "the relief of man's estate" which technology is said to enable mean except improvement in *material* conditions? While this is indubitably a good, but why is it *the* good for man? Is there greater benefit in being wealthy and powerful, or wise and virtuous? To affirm that material comforts, pleasures, and goods are the *highest* goods would directly contradict the Socratic claim on behalf of the care of the soul as the supreme concern.[7]

Aristotle likewise argues for a hierarchy of goods -the goods of the soul being most important, followed by the goods of the body, followed by external goods[8] A conception which privileges material well-being over the knowledge and cultivation of virtue (moral and intellectual) rejects this classical hierarchy which privileges the

---

[5] From *The Advancement of Learning*. Book One in) In Francis Bacon: The Major Works, 144.
[6] Idem.
[7] Plato. *The Apology*. 29e-30a.
[8] Aristotle. *Nicomachean Ethic*s. I.8.2.

good of the soul above material advantage and wealth. It is in this sense that the character of modern civilization has been called "Faustian" sacrificing the goods of the soul for material power and wealth.

References

Aquinas, St. Thomas. 1947 *Summa Theologica*. Trans. Dominican Fathers of the English Province. At DHS priory. http://dhspriory.org/thomas/summa/SS/SS182.html#SSQ182OUTP1.
Bacon, Francis. 1620 (1960). The New Organon. Fulton H. Anderson (ed.) New York: Macmillan Publishing Company
———. *In the Latin Library*. http://www.thelatinlibrary.com/bacon/bacon.liber1.shtml. Accessed May 2018.
———. 2008. Francis Bacon: *The Major Works*. Oxford: Oxfor University Press.
Hugh of St. Victor. *Erudutionis Didascalicae. Libri Septem. (Liber Secundus. XXI)*. In Documenta Catholica. http://www.documentacatholicaomnia.eu/02m/1096-1141,_Hugo_De_S_Victore,_Eruditionis_Didascalicae_Libri_Septem,_MLT.pdf. From Migne Patrologia. Accessed May 2018.
Plato. 2001 (reprint). *Euthyphro. Apology. Crito.Phaedo. Phaedrus*. Ed. Jeffrey Henderson. Loeb Classical Library, Harvard University Press.

Chapter 9
Breaking Aristotle's Bridge: The Modern Philosophical Critique of Teleology

9.1 Sir. Francis Bacon

For Aristotle the principal occupation of philosophy and thus of the theoretic life is nothing less than the knowledge of all Being through its causes and principles. The re-evaluation of causality, and in particular of final causality by modern philosophy is thus pregnant with profound consequences. A perspicacious analysis of the consequence of the modern rejection teleology for ethics and politics is found in the work of Leo Strauss, that great modern defender of the philosophical life who states plainly that:

> The rejection of final causes (and therewith also of the concept of chance) destroyed the theoretical basis of classical political philosophy.[1]

The decline of classical natural teleology in the historical-cultural sense created a vacuum which transformed modern thought entirely. Among the most important consequences are those which bear directly on the problem of the theoretic life which was conceived by Aristotle as the finality of man himself.

The claim of classical and especially Aristotelian philosophy for the theoretic life as the supreme form of human life presupposes that it is possible through reflection on human nature presupposes that first it is possible for philosophy to discern a natural hierarchy of goods (e.g. pleasure, honor, wisdom), and secondly that it is possible to discern which belong in superior and inferior positions (e.g. wisdom as a good of the soul is higher than physical pleasure considered as a good of the body.) This in turn presupposes a hierarchy of desires within the human soul which correspond to these natural goods (e.g. the intellectual desire for wisdom and the knowledge of virtue is higher than the physical desires for food and sex.) In short, the claim for the theoretic life hinges on a notion of *teleology* – that there are prescriptive

---
[1] Leo Strauss. "The Three Waves of Modernity." in *An Introduction to Political Philosophy: Ten Essays by Leo Strauss*. Hilail Gilden(ed.) (Detroit, Michigan: Wayne State University Press, 1989): 87.

purposes or ends in nature. While this idea is fully developed in Aristotle it exists in nascent form also in Plato when he speaks of patterns of conduct which are "according to nature" or "against nature." Teleology thus represents the most essential bridge between metaphysics and physics and ethics and politics. The method of Aristotle is first to show that beings have ends which correspond to and fulfill their own nature, and secondly to show that man himself has an end corresponding to his nature. This end according to Aristotle is his distinguishing function. Thus the end of man cannot consist in nutritive functions shared with plants and animals, or sensitive functions shared with animals, but only in his rational function which distinguishes him from other animals. As the theoretical activity most perfectly fulfills man's rational nature, the theoretic life will be the highest for man.

The critique of teleology in modern philosophy is therefore a vitally crucial element in undermining of the classical ideal of the theoretic life. If nothing in nature has a function, then it follows than man has no natural function. In this case it is impossible for the theoretic life to be justified as the highest mode of life by its fulfillment of the human function. And there are broader implications. Since teleology is the link between natural philosophy and ethics, the effect of the demolition of teleology is create in modern thought a radical bifurcation between the realm of natural science and that of ethics. And Leo Strauss capably argues in his *Natural Right and History* this positivistic distinction between "facts" (the ostensible domain of science) and "values" (the domain of ethics) is a fertile source of modern relativism. If natural science is the only source of genuine knowledge, and if its domain is only that of facts and not values, it follows that ethics are not a question of truth but only of preferences. How this situation arose is the topic of this chapter. We shall look at three pivotal figures central to the critique of final causality. First of course Sir. Francis Bacon, secondly Rene Descartes, and third Benedict Spinoza.

Bacon's consideration of Aristotle's Four Causes is related to his division of the sciences. He will argue that consideration of final causality does not belong to the practical as opposed the speculative science as is moreover the least fruitful of all forms of inquiry into causes within the realm of natural philosophy. As he writes with pointed eloquence:

> THE practical doctrine of nature we likewise necessarily divide into two parts, corresponding to those of speculative; for physics, or the inquiry of efficient and material causes produces mechanics, and metaphysics, the inquiry of forms, produces magic; while the inquiry of final causes is a barren thing, or as a virgin consecrated to God.[2]

The "barren" nature of teleology for Bacon may be understood in a number of senses. As we have Bacon is fundamentally interested in the question of augmenting human power which hinges on the capacity to re-produce the effects of nature for human purposes. For this project knowledge of the material and efficient causes is essential, while knowledge of final causes (apart from human intentions and purposes) is "fruitless" is the sense we have discussed – not useful. Final (and formal)

---

[2] Bacon. *The Advancement of Learning*. Third Book, Chapter V http://oll.libertyfund.org/titles/1433 (accessed January 23, 2016).

9.1 Sir. Francis Bacon

causality are assigned to metaphysics – a science which as we have seen Bacon tends to demote.

Moreover, Bacon believes that the focus on natural teleology has tended to hinder the progress of physical science by searching for something which is vaguer and more chimerical that what is manifest empirically:

> …for the treating of final causes in physics has driven out the inquiry of physical ones, and made men rest in specious and shadow causes…

which Bacon describes as like

> …remoras to the ship, that hinder the sciences from holding on their course of improvement…[3]

Looking at the history of philosophy, Bacon sees in the new direction given to natural philosophy by Plato and especially Aristotle, a step *backwards* from the mechanistic approaches to philosophy of nature found in the mechanistic (i.e. pre-teleological) approaches of Democritus who saw in nature only matter in motion.[4]

Moreover, as Bacon argues in his *Novum Organum*, the attribution of purpose to nature carries with it the danger of anthropomorphizing the natural order. The farmer who thinks it is the purpose of rain to help fertilize his fields may be attributing his own purpose rather than discerning that of nature. Seeing nature through a human perspective without objectivity hinders the progress of knowledge since "…final causes… "are clearly more allied to man's own nature, than the system of the universe, and from this source they have wonderfully corrupted philosophy."[5]

It is important however to note that Bacon neither achieves (nor even attempts) a genuine philosophical refutation of final causality. On the contrary he merely removes it from the sphere of science he happens to be interested in and which he sees as fruitful for the technological project. While he thinks the idea of final causality has hindered the advancement of physical science, he concedes the real possibility in of teleology as a genuine metaphysical reality:

> These final causes, however, are not false, or unworthy of inquiry into metaphysics, but their excursion into the limits of physical causes hath made a great devastation in that province…[6]

---

[3] Ibid. Third Book, Chapter IV – http://oll.libertyfund.org/titles/1433 (accessed January 23, 2016).
[4] Ibid.
[5] Bacon. *Novum Organum*.Aphorisms Book One, XLVIII. http://oll.libertyfund.org/titles/1432 (accessed January 23, 2016).
[6] Bacon. *The Advancement of Learning*. Third Book, Chapter IV http://oll.libertyfund.org/titles/1433 (accessed January 23, 2016).

9.2 Rene Descartes

In spite of his rationalism, it is Descartes who most decisively turns and begins to restrict reason to the domain of scientific empiricism in concert with modern mathematics which he did a considerable amount to advance. Objective reality is limited to what can be mathematically charted as quantity and magnitude (*res extensa*), while the whole qualitative realm of reality is consigned to human subjectivity (*res cogitans*).[7] What would be more accurate is that Descartes reduces reality to those elements which are serviceable for the modern project of technological domination.

Before turning then to the specifics of Rene Descartes's critique of final causality it would be well to introduce his general project so as to see where his anti-teleological position fits within the structure of his thought. Descartes has been aptly placed together with Sir. Francis Bacon among the "prophets of scientific civilization."[8] Indeed Descartes shares with Bacon the "Faustian" assumption which we have described as central to the modern turn – the idea of power as the end of knowledge. For Descartes as much as for Bacon there is a technological orientation which is seen as the key to human dominion over nature. As he writes in his *Discourse on Method*:

> For they [the principles of physics] caused me to see that it is possible to attain knowledge which is very useful in life, and that, instead of that speculative philosophy which is taught in the Schools, we may find a practical philosophy by means of which, knowing the force and the action of fire, water, air, the stars, heavens and all bodies that environ us…we can in the same way employ them to all those uses to which they are adapted, and thus render ourselves *the masters and possessors of nature.*[9]

This modern orientation tends to exalt utility. For all that Descartes remains sufficiently rooted in the classical conception of the sciences to see the need for metaphysics or "first philosophy" as a universal science which will ground all the particular science. As with Aristotle, Descartes's conception of first philosophy will include a discussion of "causes and principles" and we can therefore see how Aristotle's four causes – efficient, formal, material, and final – fit into project of establishing the grounds of science.

For Descartes it was first necessary to establish the grounds of certitude in general before one can assert knowledge claims in the particular sciences. All empirical science for example rests on sense knowledge. Hence if the validity of sense knowledge is not established the whole claim of science to knowledge is ungrounded and without certain foundation. Questions of knowledge in the history of Western

---

[7] For more on this theme see Seyyed Hossein Nasr´s Lecture "Descartes and the Fallacy of Cartesian Dualism" online http://www.bosmedia.org/musiclibrary/mp32.php?v=Tt1u6UmJ3fk (accessed February 15, 2017).

[8] R.R. Palmer and Joel Colton. 1971. *A History of the Modern World.*: 296.

[9] Rene Descartes. *Discourse on Method*. 6 in *Discourse on Method and Meditations*. Elizabeth S. Haldane and G.R.T. Haldane(trans.) (Mineola, New York: Dover Publications, 2003):41 – accessed via Google Books. (My brackets and italics.)

philosophy have indeed hinged on this issue of the validity of the senses. Aristotle and his successors among the scholastics take the senses as the starting point, and this view is radicalized by modern empiricism into a critique of the very possibility of metaphysics. Descartes belongs to the second school associated with Plato (who moderated the radicalism of Parmenides) which denies the self-evidence of the senses. Plato sees the senses as a shadowy world, a realm of "opinion" (δοξα) rather than knowledge properly speaking. Philosophers who doubt the certitude of knowledge arising from the senses are compelled to account for how universal ideas can arise or exist, since a universal idea is not instantiated in the empirical world. Hence, we find in the history of philosophy Plato's theory of recollection, Augustine´s theory of divine illumination, or Descartes's theory of innate ideas. Descartes skepticism of the senses however is more methodological than Plato´s in that his aim is to justify rather than reject the truth of what they reveal.

The findings of the Cartesian project are well known from his *Meditations*. Beginning with the maxim *de omnibus dubitandem* "That all things must be doubted" as means to discover what cannot be doubted and hence provides the grounds of certitude. All that belongs to the senses is certainly capable of being doubted – as we know from dreams that one can sense things which lack all reality. *Descartes* finds the grounds of certitude in the ego as a thinking being (*Res Cogitans*.) Following in the lines of Augustine´s *si fallor sum*, Descartes asserts that even if he is deceived, he cannot doubt his own existence as he must exit in order to doubt. Thought therefore has more self-evident reality than the physical world revealed by the senses. It is here, at this point that Descartes comes to consider the problem of causality. The ego is certain of its own existence. But how did the ego itself it come to be? It is this problem which is at least one of his routes to argue for the necessary existence of God. While Descartes is better known for a form of the ontological argument in his Third Meditation, for the sake of our discussion of causality his cosmological argument from the order of efficient causes is more immediately relevant:

> In respect of this cause one may again inquire whether it derives its existence from itself or from another cause. If from itself, then it is clear from what has been said that it is itself God, since if it has the power of existing through its own might, then undoubtedly it also has the power of actually possessing all the perfections of which it has an idea – that is, all the perfections which I conceive to be in God. If, on the other hand, it derives its existence from another cause, then the same question may be repeated concerning this further cause, namely whether it derives its existence from itself or from another cause, until eventually the ultimate cause is reached, and this will be God."[10]

Descartes's language makes it sound as if God is His own Cause – which brought forth the objections of the scholastics. While Thomists allowed for the idea of God as *ipsum esse subsistens* (self-subsistent Being), they did not allow for the idea of

---

[10] From Descartes, *Meditations III*. Quoted in Daniel E. Flage and Clarence A. Bonnen. "Descartes on Causation". In *Review of Metaphysics*. This seems to be the John Cottingham translation - quote crossed checked in First Philosophy Fundamental Readings in Philosophy.II Knowledge and Reality. (Andrew Bailey (ed.) (Broadview Press, 2011). https://books.google.es/books?id=0hJbD wAAQBAJ&printsec=frontcover#v=onepage&q&f=false (accessed 10/27/18)

God as *causa sui* (self-caused), for this would make God into an effect, and furthermore the cause is presumptively prior to the effect which according to Aristotle is an impossibility. It is on this account that Descartes refers to the distinction between the formal and efficient causes. God cannot be the efficient cause of Himself, but He can be the formal cause in the sense that that God exists through His essence. This issue arises in response to the objection by the scholastic Antoine Arnauld:

> …the question ´why does God exist 'should be answered not in terms of an efficient in the strict sense, but simply in terms of the essence or formal cause of the thing.[11]

With respect to material causes or as Descartes would prefer substance or things (res), the demonstration of God's existence provides the ground of certitude for the correspondence of sense impressions to real bodies (i.e. since God is not a deceiver.) However Descartes will argue that its real qualities are the t qualitative elements perceived by the senses – as color or hardness, *sed tantum in eo, quod sit res extensa in longum, latum, &profundum* "…but only inasmuch as it is an extended thing in length, width, and depth."[12]

Descartes quantification of material substances is naturally of vital importance to the scientific revolution in that provided philosophical grounds for the much discussed "mathematization of nature" – i.e. the idea of mathematics as the language which renders the physical world rationally intelligible. Clearly this process was involved in the progress of physics and astronomy in the great works of Kepler, Galileo, and Newton. Descartes own contributions to mathematics – perhaps most famously analytic geometry – are to the practical advancement of the sciences.

We have now seen for all Descartes's radicalism, that the Aristotelian concepts of efficient, formal, and material causes all find a place within his system or project. Only final causality poses a fundamental problem for Descartes. The reason is centrally because Descartes conceives the idea of *purpose* as belonging to the divine Creator whose ultimate purposes are incomprehensible. As he says in the Fourth Meditation:

> …I have no longer any difficulty in discerning that there is an infinity of things in his power whose causes transcend the grasp of my mind: and this consideration alone is sufficient to convince me, that the whole class of final causes is of no avail in physical [or natural] things; for it appears to me that I cannot, without exposing myself to the charge of temerity, seek to discover the [impenetrable] ends of Deity.[13]

The same claim that the quest for final causes involves a "presumption" – an effort to discern incomprehensible divine purposes appears also in his *Principles of Philosophy.*

---

[11] Descartes in his reply to the Fourth Objection (Arnauld) in http://www.earlymoderntexts.com/assets/pdfs/descartes1642_2.pdf (accessed 2/20/16).

[12] Rene Descartes. *Principia Philosophiae* 2:4 Latin text at https://books.google.es/books?id=IHpbAAAAQAAJ&redir_esc=y (accessed 2/20/16). The translation was mine.

[13] Rene Descartes. *Meditations on First Philosophy.* IV.6 in http://www.wright.edu/~charles.taylor/descartes/meditation4.html (accessed 2/28/16) – words in the brackets are in original text.

Likewise, finally, we will not seek reasons of natural things from the end which God or nature proposed to himself in their creation (i.e. "final causes), for we ought not to presume so far as to share in the counsels of Deity, but considering him as the efficient cause of all things, applied to some of his attributes we should have some knowledge...[14]

Note that this argument does not even engage Aristotle's argument for an *immanent* teleology knowable from the operations of nature. It *asserts* that purpose in nature can only exist as the hidden purposes of God which are not manifested empirically. In the absence of knowable, intrinsic purposes within the natural order to which man must conform, the way is open for man to assert *his* purposes upon a natural order he subordinates to *his* ends.

9.3 Benedict Spinoza

The third and most strident and substantive critique of final causality is found in the work of Benedict Spinoza. Unlike Bacon and Descartes, Spinoza aims not merely to sideline final causes (Bacon) or consign them to the incomprehensibility of the divine mind (Descartes), but to disprove their logical possibility altogether:

> There will now be no need of many words to show that nature has set no end before herself, and that all final causes are nothing but human fictions.[15]

Spinoza's rejection of final causality stems largely from the peculiarities of his necessitarian and semi-pantheistic theology. Whereas Descartes consigned final causes to the incomprehensible divine mind, Spinoza claims sufficient comprehension to know that… "all things are begotten by a certain eternal necessity of nature and in absolute perfection."[16] Spinoza thinks he has probative reasons to believe God does not act by free will but only through necessity (Proposition 32, Corollary I), a fact which Spinoza takes to rule out the possibility of *purposeful* action. Precisely why this is so being not elaborated, but we may presume that for Spinoza the idea of purpose requires a faculty of choice and deliberation.

Moreover, according to Spinoza, the idea of final causality conflicts with his idea of the order of perfection in nature. Spinoza argues that the most perfect effect will be the one most *immediately* produced by God while final causality implies the greater perfection of the last in a chain of inter-mediate causes.[17]

Moreover, he argues that final causality suggests insufficiency in God this contradicting the divine Perfection – "…if God works to obtain an end, He necessarily seeks something of which he stands in need."[18]

---

[14] Rene Descartes. *Principles of Philosophy*. I.28 – (Radford, VA: Wilder Publications, 2008): 28.
[15] Benedict Spinoza. *Ethics*. I. (Appendix) translation by W.W. White, revised by A.H. Stirling (Ware, Herfordshire: Wordsworth, UK): 37.
[16] Idem.
[17] Idem.
[18] Ibid. 38.

Spinoza's critique of final causality hinges on his peculiar theology – a theology which of course Judaism and Christianity would reject. God is held to act and create freely, out of a benignant desire to share His Goodness with His creatures. But what is curious is how radically Spinoza like Descartes before him "theologizes" the whole question of final causality. For Aristotle himself as we have seen one teleology is inferred from the character of the actual operations of nature (e.g. *Physics* II.8). Of course, natural teleology may have theological implications – implications drawn out with enthusiasm by the medieval scholastic philosophers. But Aristotle gives no indication that the truth of natural teleology hinges on any such claimed insight into the divine mind as Descartes and Spinoza assume in their ostensible "refutations."

Another element of Spinoza's critique of final causality which he shares with Bacon is his belief that final causality is an anthropomorphic projection of human notions of purpose on to nature. As he writes:

> The attempt, however, to show that nature does nothing in vain (that is nothing which is not profitable to man), seems to end in showing that nature, the gods, and man are alike mad… not a few things must have been observed which are injurious, such as storms, earthquakes, diseases, and it was affirmed that these thing happened because the gods were angry because of wrongs which had been inflicted on them by man, or because of sins committed in the method of worshiping them…[19]

Taken by itself as an argument against final causality, this position rests on a genetic fallacy. Simply because human beings tend to project final causality on to nature does not prove that there is no final causality in nature. Indeed, the consonance between the tendencies of the human mind and those of nature might well be deemed "fitting". And while one can accept that the Baconian and Spinozistic cautions are valuable in helping the observer of nature guard against projecting too easily human notions of purpose on to the natural order, they don't prove that purpose is universally an invalid category when applied to the natural order.

The Ethical-Political Implications of Modern Anti-Teleology

Leo Strauss argues for a fundamental distinction in the approach of Plato and Aristotle:

> Plato never discusses any subject…without keeping in view the elementary Socratic question, 'what is the right way of life?' '…Aristotle on the other hand, treats of the various levels of beings, and hence especially every level of human life, on its own terms.[20]

---

[19] Ibid. 36–37.
[20] From *Natural Right and History, (1953),* 156 quoted in John P. East "Escaping the Stifling Clutches of Historicism" in http://www.theimaginativeconservative.org/2016/04/leo-strauss-escaping-the-stifling-clutches-of-historicism.html (accessed 4/29/2016).

9.3 Benedict Spinoza

There is apparent truth here in that Aristotle's natural and speculative philosophy is not as evidently and immediately tied to ethics as ae nearly all the concerns of the Platonic Socrates. And yet we must consider that for Aristotle the theoretic life which embraces the speculative and scientific concerns of metaphysics and natural philosophy *is* the "right way of life". Furthermore, the theoretic life as Aristotle's answer to the Socratic question "what is the good life?" hinges on his ideas of natural teleology which furnishes the basis of fields as seemingly remote from each other as physics and politics. Only if there are ends in nature is it possible to claim that man himself has a natural end, and this end will be the supreme Good which orders man's ethical and political life. Teleology therefore is *the bridge* between nature and ethics.

The ostensible destruction of the natural teleology by modern philosophy led to the construction of a *new science of nature*. This modern form of science is one which excludes the final and formal causes from the purview of empirical investigation and restricts itself to the consideration of material and efficient causes. The consequent loss of the bridge leading from nature to ethics is obviously pregnant with the most significant implications for ethical and political philosophy. The figure who did the most to first work out these implications with steadfast consistency and to bring the Baconian revolution into the world of man, was Thomas Hobbes.

The direct contradiction between Aristotle and Hobbes can be most clearly seen on the question of whether or not there exists a Supreme Good (τἀγαθὸν καὶ τὸ ἄριστον) of human striving. For Aristotle the existence of such a good is logically required by the teleological moral structure of human actions. All human acts aim at the possession of some good, though a human act may aim (and most often *does* aim) for the possession of a good which is merely useful for the possession of a good beyond itself (as exercise or medicine to health.). The existence of intrinsic goods however logically necessitates the existence of Supreme Good for otherwise no action would have any ultimate goal. We will recall Aristotle's argument that:

> If therefore among the ends at which our actions aim there be one which we wish for its own sake, while we wish others only for the sake of this, and if we do not choose everything for the sake of something else (which would logically result in a process *ad infinitum*, so that all desire would be futile and vain), it is clear that this one ultimate End must be the Good, and the Supreme Good.[21]

This concept of the supreme good which will be the key to Aristotelian ethics and politics, is rejected directly and out of hand by Hobbes:

> ...we are to consider that the felicity of this life consisteth not in the repose of a mind satisfied. For there is no finis ultimis (utmost aim) no summum bonum (greatest good) as is spoken of in the books of the old moral philosophers. Nor can a man any more live whose desires are at an end than he whose senses and imaginations are at a stand. Felicity is a continual progress of the desire from one object to another, the attaining of the former being still but the way to the latter...the voluntary actions and inclinations of all men tend not only to the procuring, but also the assuring of a contented life, and differ only in the way which

---

[21] Aristotle. *Nicomachean Ethics* (H. Rackham) I.ii.1–2.

ariseth partly from the passions in diverse men, and partly from the difference of the knowledge or opinion which each one has of the causes which produce the effect desired.[22]

Hobbes psychological portrait of human life here recalls that of Buddhism or Schopenhauer –the whole of man's existence is one characterized by a restless and unending movement from one desire to the next without the possibility of quiescence. The fulfillment of one desire simply opens the door to a new one. Moreover, whereas Aristotle considers the Supreme Good which orders human desire to be the same for all humanity, Hobbes emphasizes the individual variability of desire arising from the diversity of the passions so that "the Good" is not a concept capable of a univocal definition for all men. Hobbes is ready then to admit what Aristotle could not – the ultimately aimlessness and vanity of human desires.

Aristotle of course was also aware of the *de facto* diversity of ends which men pursue, and the consequent variations in individual definitions of "the Good." However, as Leo Strauss notes, this fact so far from disproving the existence of *"the* Good" is the condition that motivates the philosophical quest itself. Let us assume that some conceptions of the Good are more correct than others. In that case the brute fact attested by observation that men differ in their notions of the Good, provokes the question of which if any Good is the *true* Good, and which "goods" are merely apparent. And how can they be distinguished? Here the teleological structure of nature intervenes to resolve the difficulty, for man like all other beings in nature has a unique End determined by his essential nature and place in the natural hierarchy – in this case as a *rational* animal. Those who seek for instance the Supreme Good in bodily pleasures are deceived, for the pleasures of food and sex are shared with the other animals. Aristotle's definition of the good life as "…the active exercise of the soul´s faculties in conformity with rational principle…"[23] satisfies the requirement of the true Good, because it is a while the higher activities of the intellect are a distinguishing function of man. Though not all attain to the full expression of this form of life (the theoretic life), it can be nonetheless said that the life of reason is nonetheless vindicated as the most perfect fulfillment of man universally – i.e. *qua* man.

Since Hobbes rejects natural teleology, and since without a Supreme Good, there seems to way to establish a hierarchy among the goods which men desire, there seems to be no way to distinguish true goods from merely apparent ones. The practical variability of human ends, and the inability to order them by any objective standard becomes simply a basic reality with which political philosophy must reckon.

For Aristotle since all mankind shares a *common* Supreme Good, it follows that the purpose of the political order must transcend mere physical security and include also the cultivation not merely of these *means* of life, but also of the *end* of life – i.e. the *good* life:

---

[22] Thomas Hobbes. *Leviathan*. XI. in http://www.ttu.ee/public/m/mart-murdvee/EconPsy/6/Hobbes_Thomas_1660_The_Leviathan.pdf (accessed 5/3/2016). Slight differences with text in Penguin,1985 (C.B. Macpherson, ed.)

[23] Aristotle. *Nicomachean Ethics*.I.8.14.

It is manifest therefore that the state is not merely the sharing of a common locality for the purpose of preventing mutual injury and exchanging goods. These are necessary preconditions of a state's existence, yet nevertheless, even if all these conditions are present, that does not therefore make a state…the object of a state is the good life, these things are means to that end.[24]

If the end of politics is the good life, and if the good life embraces not merely the moral but also intellectual excellence, it follows then that the theoretical life as the most perfect fulfillment of the intellectual virtues will be part of the end of the political life.

For Hobbes and generally speaking the whole modern contractarian tradition from the seventeenth century onward, the basis and end of the state is precisely what Aristotle says it is not -an agreement for the purpose of "avoiding mutual injury and exchanging goods." Since there is no natural Supreme Good or End for mankind, nor anyway of coming to agreement on these points, hence the Summum Bonum cannot be the proper end of politics. There is however agreement on what Aristotle consigns to the mere means to the attainment of the good life, namely those things needed for the *preservation* of life at all. As Hobbes says

THE right of nature, which writers commonly call jus naturale, is the liberty each man hath to use his own power as he will himself for the preservation of his own nature…[25]

Since his self-preservation is in jeopardy from the absence of security in the state of nature, nature drives man to seek the condition of peace through the social contract which establishes the political society.

This central move in the thought of Hobbes thus leads directly to what Leo Strauss terms the "lowering of sights" in modern political philosophy from excellence to mere existence. Where Aristotle sees not life but the *good* life, as the end of political society Hobbes merely sees life; what for Aristotle is merely a means for Hobbes is the end.

In spite of the authoritarianism of Hobbes, Strauss astutely judges his him as a progenitor of modern liberalism.[26] This is largely because by removing the problem of "the good life" from the sphere of politics, Hobbes's theory opens the door to a pragmatic problems which plagued the 16th and 17th centuries with their religious wars and civil strife – how to make possible the co-existence of different conceptions of the good within a common polity. Hobbes´s "resolves" the question of the good by annulling it, or at best consigning it to the n private individual sphere. The presumption is that while conceptions of the good life are a cause of dispute, the

---

[24] Aristotle. *Politics*. Book II.1280b in http://www.perseus.tufts.edu/hopper/text?doc=Perseus%3A text%3A1999.01.0058%3Abook%3D3%3Asection%3D1280b (accessed 5/22/16) – see Miller, Fred, "Aristotle's Political Theory", *The Stanford Encyclopedia of Philosophy* (-->Fall 2012 Edition), Edward N. Zalta (ed.), URL = http://plato.stanford.edu/archives/fall2012/entries/aristotle-politics/ (5/22/2016).

[25] Thomas Hobbes. *Leviathan*. Part I, Chapter XIV -www.ttu.cc/public/m/mart murdvee/…/ Hobbes_Thomas_1660_The_Leviathan.pdf (accessed 5/22/2016).

[26] Leo Strauss. *Natural Right and History*. (University of Chicago Press, Copyright 1953) – 181–182.

goodness of living is a topic of broad consensus. This presumes one imagines a basic skepticism about the capacity of man to know the nature of the good life – a conception which no doubt follows not simply from philosophical sloth but from the Baconian-Hobbesian rejection of natural teleology which we have endeavored to outline. A further natural consequence of this movement of thought is the eclipse of the theoretic life as the human ideal.

References

Aristotle. 1926 (1999 reprint). *Nicomachean Ethics*. 1926. (1999 reprint). *Nicomachean Ethics*. Trans. H. Rackham, 1999 Loeb Classical Library, Harvard University Press.
———. 1932 (2005 reprint). *Politics*. Trans. H. Rackham. Loeb Classical Library, Harvard University Press Sometimes accessed at Tufts via Perseus – http://www.perseus.tufts.edu/hopper/text?doc=Perseus%3Atext%3A1999.01.0058%3Abook%3D3%3Asection%3D1280b. Accessed 22 May 2016.
Bacon, Francis. *The Advancement of Learning*. Liberty Fund. http://oll.libertyfund.org/titles/1433. Accessed 23 Jan 2016.
———. *Novum Organum*. "Aphorisms." Liberty Fund. http://oll.libertyfund.org/titles/1432. Accessed 23 Jan 2016.
Descartes, Rene. 2003. *Discourse on Method and Meditations*. Trans. Elizabeth S. Haldane and G.R.T. Haldane. Mineola: Dover Publications. Accessed 28 Feb 2016.
———. 2008. *Principles of Philosophy*. Radford: Wilder Publications.
———. *Principia Philosophiae* Latin text at https://books.google.es/books?id=lHpbAAAAQAAJ&redir_esc=y. Accessed 20 Feb 2016.
———. Reply to the Fourth Objection (of Arnauld). in http://www.earlymoderntexts.com/assets/pdfs/descartes1642_2.pdf. Accessed 20 Feb 2016.
———. *Meditations on First Philosophy*. At Wright.edu. http://www.wright.edu/~charles.taylor/descartes/meditation4.html. Accessed 28 Feb 2016.
East, John P. Escaping the Stifling Clutches of Historicism. *At the Imaginative Conservative*. http://www.theimaginativeconservative.org/2016/04/leo-strauss-escaping-the-stifling-clutches-of-historicism.html. Accessed 29 Apr 2016.
Hobbes, Thomas. 1660. *The Leviathan*. At TTU. http://www.ttu.ee/public/m/mart-murdvee/EconPsy/6/Hobbes_Thomas_1660_The_Leviathan.pdf. Accessed 3 May 2016.
Miller, Fred. 2012. Aristotle's Political Theory. In *At the Stanford Encyclopedia of Philosophy*. (Fall 2012 Edition), ed. Edward N. Zalta. URL = http://plato.stanford.edu/archives/fall2012/entries/aristotle-politics/. Accessed 22 May 2016.
Nasr, Seyyed Hossein. Lecture. Descartes and the Fallacy of Cartesian Dualism online. http://www.bosmedia.org/musiclibrary/mp32.php?v=Tt1u6UmJ3fk. Accessed 15 Feb 2017.
Palmer, R.R., and Joel Colton. 1971. *A History of the Modern World*. 4th ed. New York: Alfred A Knopf.
Spinoza, Benedict. 2001 *Ethics*. I. Trans. W.W. White, Revised by A.H. Stirling. Ware: Wordsworth.
Strauss, Leo. 1953 (1999 reprint). *Natural Right and History*. Chicago: University of Chicago Press.
———. 1989. The Three Waves of Modernity. In *An Introduction to Political Philosophy: Ten Essays by Leo Strauss*, ed. Hilail Gilden. Detroit: Wayne State University Press.

Chapter 10
The Enlightenment as a Baconian Revolution

> *"It must be admitted that the inventors of the mechanical arts have been much more useful to mankind than the inventors of syllogisms..."(Voltaire. Philosophical Dictionary.) https://history.hanover.edu/texts/voltaire/volpreci.html (accessed May 13, 2018) translated by H.I. Woolf: Knopf, 1924 – scanned by Hanover in 1995. elected and 4) -Voltaire. Philosophical Dictionary (Precis of Ancient Philosophy)*

The Enlightenment of the late seventeenth and eighteenth centuries is often described as the "Age of Reason." Yet as a philosophical movement culminating in Hume and Kant it was better characterized by Pope Benedict XVI when he spoke of "…the modern *self-limitation* of reason…"[1] -that is to say the limitation of reason to the empirical world charted by the physical sciences. Perhaps then we may say that the Enlightenment is about the replacement of *metaphysical* rationality with an *empirical* and *instrumental* rationality –scientific knowledge at the service of the technological project. It can be said in general that the Enlightenment was the movement which carried forward the Baconian revolution and eroded the underpinnings of the classical theoretic ideal. The connection between Bacon and the Enlightenment was already identified clearly by its fiercest critic Count Joseph de Maistre.[2]

Baconianism had already played a leading role in the institutionalization of modern science in England. In 1660 in Gresham college, London a group of devotees of Bacon's approach to natural philosophy including such luminaries as the chemist Robert Boyle and the polymath Robert Hooke formed a society for the promoting

---

[1] Pope Benedict XVI. *Faith, Reason, and the University: Memories and Reflections* often called "The Regensburg Lecture" (2006) http://w2.vatican.va/content/benedict-xvi/en/speeches/2006/september/documents/hf_ben_xvi_spe_20060912_university-regensburg.html accessed (June 16, 2016). My ital.

[2] De Maistre wrote a work entitled *Examination of the Philosophy of Bacon*.

of "physico-mathematical experimental learning."³ In 1662 it received a charter as The Royal Society from King Charles II which became the leading promotor of the natural sciences in England. The Royal Society reached its crowning achievement with the publication of Sir. Isaac Newton's *Principia Mathematica* in 1687. By synthesizing Galileo's principles of terrestrial motion and Kepler's principles of planetary motion into a uniform theory of motion Newton brings the entirety of the scientific-revolution to a glorious culmination. The work of the Royal Society culminating in the Newtonian revolution seems like the full vindication of Bacon's inductive methodologies in concert with modern developments in mathematics, as well as his advocacy for collaboration in the sciences. Man learned to understand the principles of nature in their elegant mathematical exactitude and simplicity. And in so doing began to gain power over it, as knowledge of physics enabled technology. It is fitting that Thomas Sprat in his 1667 frontispiece of his *The History of the Royal Society of London* places Bacon alongside its patron Charles II. The establishment of the Royal Society was soon followed by the creation of other scientific societies in continental Europe.

The institutionalization of the Baconian Revolution was carried forward by the central ideologists of the Enlightenment, the *philosophes* of France. The leading figure of their number Voltaire was instrumental in popularizing the works of Bacon to the literate French public (if not indeed throughout the Western world.) In his *Letters on the English* (1778) Voltaire writes of Bacon that while slow and relatively unknown improvement in empirical science and technology occurred before Bacon, nonetheless:

> He is the father of experimental philosophy…no one before Lord Bacon was acquainted with experimental philosophy, nor with the several physical experiments which have been made since his time…In a little time experimental philosophy began to be cultivated on a sudden in most parts of Europe. It was a hidden treasure which the Lord Bacon had some notion of, and which all the philosophers, encouraged by his promises, endeavoured to dig up.⁴

The defining text of the eighteenth century Enlightenment, the French Encyclopedia edited by Denis Diderot and Jean D'Álembert give to Francis Bacon a position which approaches religious veneration. They are beyond lavish in the praises of him, and entirely accept Bacon's conceit to have renovated the sciences out of a period of darkness:

> The immortal Chancellor of England, Francis Bacon (1561–1626), ought to be placed at the head of these illustrious personages. His works, so justly esteemed (and more esteemed, indeed, than they are known), merit out reading even more than out praises (19). One would be tempted to regard him as the greatest, the most universal, and the most eloquent of the philosophers, considering his sound and broad views, the multitude of objects to which his mind turned itself, and the boldness of his style, which everywhere joined the most sublime

---

³There is good information on the history of the Royal Society at their website. https://royalsociety.org/about-us/history/ (last accessed May 2018.)

⁴Voltaire. *Letters on the English* in https://legacy.fordham.edu/halsall/mod/1778voltaire-lettres.asp (accessed June 7, 2016). Also https://www.bartleby.com/34/2/12.html (October 27, 2018)

images with the most rigorous precision. Born in the depths of the most profound night, Bacon was aware that philosophy did not yet exist...[5]

The Encyclopedists were of course in an intellectual struggle with the Roman Catholic Church and saw Bacon as useful for breaking with the system of Catholic scholasticism, which was primarily a synthesis of Christian theology and Aristotelianism. Scholasticism is referred to in the Encyclopedia as "the science of the centuries of ignorance."[6]

Bacon is thus especially credited with shaking off the influence of Catholic scholasticism and suggests that his project could be pushed forward still further.

> Scholasticism, which continued to dominate, could not be overthrown except by bold and new opinions...if we did not know with what discretion, and with what superstition almost, one ought to judge a genius so sublime, we might even dare reproach Chancellor Bacon for having perhaps been too timid. He asserted that the scholastics enervated science with their petty questions and that the mind ought to sacrifice the study of general beings for that of individual objects; nonetheless he seems to have shown a little too much caution or deference to the dominant taste of his century in his frequent use of the terms of the scholastics...[7]

Of special importance is the fact that the so called "Tree of Diderot and D'Álembert", the taxonomical classification of the sciences in the Encyclopedia is directly adapted from Bacon's schema.[8] In other words the modern understanding of thinking about the sciences, their hierarchy, and inter-relationships is essentially Baconian. This represented a remarkable shift in French intellectual life from the strict rationalism of the Cartesian school to scientific empiricism (though retaining the mathematical orientation of Descartes.)

It is a measure of Bacon's importance that moreover the culminating of Enlightenment thought, Immanuel Kant opens his *Critique of Pure Reason* with a quotation from the preface of Bacon's *Instauratio Magna* and goes on to remark:

> It took natural science longer to find the highway of science; for it is only one and half centuries since the suggestion of the ingenious Francis Bacon partly occasioned this discovery and partly further stimulated it...[9]

The basic Baconianism of the Enlightenment epoch went hand in hand with the devaluation of the theoretical life as the human ideal. Leo Strauss sees Jean Jacques Rousseau as a pivotal figure in this respect, and this on three grounds two of them more "social" than the Aristotelian conception and one more individualistic. First, Rousseau privileges practical assistance to one's neighbor as the only requisite to

---

[5] Diderot and DÁlembert. *Preliminary Discourse to the Encyclopedia of Diderot*. Translated by Richard Schwab with assistance of Walter E. Rex. http://quod.lib.umich.edu/cgi/t/text/text-idx?c=did;cc=did;rgn=main;view=text;idno=did2222.0001.083 (accessed 5/30/2016). For another text with effusive praise of Bacon see Jean DÁlembert's "Reflections on the Present State of the Republic of Letters"(1760) in *The Portable Enlightenment Reader*. Isaac Kramnick(ed.) U.S.A., Penguin, 7–17.

[6] Idem.

[7] Idem.

[8] This is stated explicitly in the preface.

[9] Immanuel Kant. *Critique of Pure Reason*, bxii (Preface). Translators/Editors Paul Guyer and Allen W. Wood. (New York: Cambridge University Press, 2007 printing):108.

the good life; secondly, he privileges patriotism – service to the community; and third he emphasizes an ideal of "self-realization."[10] While the idea of the supremacy of the "political" life has ancient roots, the ideal of "individual self-realization" is peculiarly modern. Aristotle has assumed a universal species essence – that of rationality – such that only one kind of life, the life of reason, is proper to man. Man can choose to live in another way, irrationally for instance, but such is not a properly human existence (as in the phrase "to live like an animal").

From its Baconian roots, the Enlightenment's project arrives at a broad-based critique of the entire classical ideal of the theoretic life. The social and political changes it encouraged, and their impact on the theoretic ideal will be taken up later. First, we will consider a development in Enlightenment thought of a purely intellectual character - the *critique of metaphysics*. The Enlightenment moved toward supplanting classical metaphysics with scientific empiricism. For Aristotle as we have seen metaphysics (First Philosophy") was the highest occupation of the theoretic life, the ruling science whose excellence transcended all utilitarian concerns. Thus, these developments signal a fundamental shift.

The Enlightenment was intellectual movement extraordinary in its complexity and manifold effects. It is not an exaggeration to say that the modern culture of the West in its fundamental values and ideas – liberal democracy, tolerance, egalitarianism, secularization, the free market, and the scientific-technological world view – is underpinned principally by its tenets. The social aspects of this revolution will be considered later. First, we will consider in more detail the foundations of modernity's dethronement of classical metaphysics and coronation of empirical science.

References

Benedict XVI. 2006. Benedict XVI. "Faith, Reason, and the University: Memories and Reflections" often called "The Regensburg Lecture". At the Vatican. http://w2.vatican.va/content/benedict-xvi/en/speeches/2006/september/documents/hf_ben-xvi_spe_20060912_university-regensburg.html. Accessed 16 June 2016.

D'Álembert, Jean. 1760. Reflections on the Present State of the Republic of Letters. In 1995. *The portable enlightenment reader*, ed. Isaac Kramnick. New York: Penguin.

Diderot, Denis and Jean le Rond D'Alembert. 1995 (translation of 1751 work.) *Preliminary Discourse to the Encyclopedia of Diderot*. Trans. Richard Schwab with assistance of Walter E. Rex. http://quod.lib.umich.edu/cgi/t/text/text-idx?c=did;cc=did;rgn=main;view=text;idno=did2222.0001.083. Accessed 30 May 2016.

History of the Royal Society. https://royalsociety.org/about-us/history/. Accessed May 2018.

Kant, Immanuel. 2007. *Critique of Pure Reason*. Trans. Paul Guyer and Allen W. Wood. New York: Cambridge University Press.

Strauss, Leo. 1962. Lecture transcription -"Rousseau", Autumn. https://leostrausscenter.uchicago.edu/sites/default/files/Rousseau%201962.pdf. Accessed 8 June 2016.

Voltaire. 1778. *Letters on the English*. At Fordham. From the Harvard Classics Translation (1910). At Fordham. https://legacy.fordham.edu/halsall/mod/1778voltaire-lettres.asp. Accessed 7 June 2016.

---

[10] Leo Strauss. Lecture Transcription ("Rousseau", Autumn, 1962) (https://leostrausscenter.uchicago.edu/sites/default/files/Rousseau%201962.pdf): (accessed June 8, 2016):17.

Chapter 11
Metaphysics Dethroned: Hume, Kant, and the "Self-Limitation of Reason"

> There was a time when metaphysics was indeed the queen of all the sciences…Now in accordance with the fashion of the age, the queen proves despised on all sides; and the matron, outcast and forsaken, mourns like Hecuba… –Immanuel Kant. *Critique of Pure Reason. A.ix*
>
> Why are we here? Where do we come from? Traditionally, these are questions for philosophy, but philosophy is dead…scientists have become the bearers of the torch of discovery in our quest for knowledge. -Stephen Hawking. (Google Conference, 2011)

We have earlier seen Aristotle's long lived image of first philosophy or metaphysics as the master of all the other sciences whose role is to serve it as the noblest and most universal science. To continue the metaphor the modern age has seen revolution by those who were meant to serve it – the empirical, particular sciences. These have dethroned and exiled the master science and have indeed usurped the name of science without qualification. For if Bacon challenged the *utility* of metaphysics, the leading thinkers of the Enlightenment took the next step and challenged its very *possibility*. The German philosopher G.W.F. Hegel wrote that:

> …the eclaircissement [Enlightenment]utterly banished and extirpated all that was speculative from things human and divine.[1]

Since metaphysics is the highest if not the principal object of the theoretic life in the classical conception, one can say that the modern claim against the theoretic life here reaches a new level.

This revolutionary shift in the position of speculative philosophy which was pithily described by Pope Benedict XVI as "…the modern self-limitation of reason…"[2] was well under way by the eighteenth century. It is interesting to contrast the

---

[1] G.W.F. Hegel. *The Philosophy of History*. J.Sibree (trans.) in http://socserv2.socsci.mcmaster.ca/~econ/ugcm/3ll3/hegel/history.pdf (Accessed June 15, 2016). (My brackets).

[2] Pope Benedict XVI. *Faith, Reason, and the University: Memories and Reflections* often called "The Regensburg Lecture" (2006) http://w2.vatican.va/content/benedict-xvi/en/speeches/2006/september/documents/hf_ben-xvi_spe_20060912_university-regensburg.html accessed (June 16, 2016).

position of metaphysics in the hierarchy of the sciences between the medieval period and that of Kant. We have seen Aristotle's notion of First Philosophy as the master science ruling the others. The metaphor is carried forward by St. Thomas Aquinas who writes that there must a ruling science which is the *mistress of the others:*

> Thus the philosopher teaches in his *Politics*, that when many things are ordered to one, one must be regulating or ruling, and the others must be ruled…now all of the sciences and arts are ordered to one thing, which is happiness. Whence it is necessary that one should be the mistress of the all the others….[3]

Metaphysics for Aquinas is clearly the mistress of the other sciences for:

> …therefore that science ought naturally to be the mistress of the others which is the most intellectual. That is that one which is turned most to the consideration of intelligible things.[4]

Aquinas goes on to argue that since the most complete and certain knowledge of intelligible things is that which includes the knowledge of things through their causes, and the science which considers these things is metaphysics, it is evident that metaphysics is the mistress of the sciences.

Immanuel Kant in the preface to his own magnum opus, *The Critique of Pure Reason* boldly proclaims the reign of this mistress at an end:

> There was a time when metaphysics was indeed the queen of all the sciences, and if the will be taken for the deed, it deserved this title of honor, on account of the preeminent importance of its object. Now in accordance with the fashion of the age, the queen proves despised on all sides; and the matron, outcast and forsaken, mourns like Hecuba: *modo maxima rerum, tot generis natisque potens -nunc trahor exul, inops*--Ovid. *Metamorphoses.*[5] [["Greatest of all by race and birth, I am now cast out, powerless."]]

The modern dethronement of metaphysics is ironic because modern philosophy begins with a return to the Aristotelian idea of metaphysics as the universal science. While Descartes challenged the Aristotelian system, nonetheless he aims at the same thing as Aristotle –he only wishes to reconstitute this universal science which will provide the foundations for the particular sciences on the ground of absolute

---

[3] Sicut docet philosophus in politicis suis, quando aliqua plura ordinantur ad unum, opportet unum esse regulans, sive regens, et alia regulata…omnes autem scientiae et artes ordinantur in unum, quae est. eius beatitude. Unde necesse est. quod una earum sit aliarum omnium rectrix, quae nomen sapientiae recte vindicate. Nam sapientis est. alios ordinare.St. Thomas Aquinas. *Commentary on Aristotle's Metaphysics*. The Latin text can be found here with John P. Rowan's 1961 translation http://dhspriory.org/thomas/Metaphysics.htm (accessed June 16, 2016). I checked my English translation against both Rowan's translation which I also have as a separate text *Commentary on Aristotle's Metaphysics* (Notre Dame, Indiana: Dumb Ox Books, 1995 revised edtion).

[4] ita scientia debet naturaliter aliarum regulatrix quae maxime intellectualis est. Haec autem est., quae circa maxime intelligbilia versatur. Ibid.

[5] Immanuel Kant. *Critique of Pure Reason.*Translated by Paul Guyer and Allen W. Wood. (New York: Cambridge University Press, 2007): 99 (preface A ix)The Ovid translation is from the translators and I placed it in brackets. Future references to this text will be designated CPR followed by the page number as well as the original pagination of the A or B editions. The translation of Ovid is from the text, though including it in the brackets is mine.

certitude. Indeed, he returns to Aristotle's very terminology in his canonical *Meditations on First Philosophy*.

Yet in the long arc of modern history it is plain that the Cartesian project to regenerate metaphysics to ground the modern sciences has proved abortive. While two of Aristotle's theoretical sciences – mathematics and natural philosophy (what is now called natural science) underwent prodigious development in the modern era, the principal science metaphysics stagnated and indeed its very possibility was thrown into profound question. The result is a *disintegrated* state in which many specialized sciences advance on their own terms, but there is no encyclopedic project to bring all the sciences under a unified system of knowledge. The main reasons for the lack of advancement in metaphysics are twofold. First, the acceptance of the Baconian project means that knowledge is considered from the standpoint of technical utility. Even research in the "pure" natural sciences and mathematics has been shown to ultimately have important applicability. Hence, while the empirical sciences are highly valued for their role in technology, metaphysics has been progressively devalued. A recent example of in this regard was provided by the eminent Cambridge physicist Stephen Hawking:

> Why are we here? Where do we come from? Traditionally, these are questions for philosophy, but philosophy is dead…scientists have become the bearers of the torch of discovery in our quest for knowledge.[6]

It is difficult to know whether Hawking's statement was an expression of scientific triumphalism or rather a lament. For it is quite apparent that philosophy so far from advancing with the "particular sciences" instead became caught up in a philosophical *self-critique* which gradually undermined the whole project of philosophy as the universal science. When we turn to the major figures of post-Cartesian Enlightenment philosophy we find not an advance or affirmation of the value of metaphysics but a series of sustained critiques. In this regard two figures stand out – David Hume and Immanuel Kant –both of whom can be ironically considered modern practitioners of the theoretic life, even as they are history's most brilliant critics of the possibility of metaphysics.

11.1 David Hume's Skeptical Empiricism

Classical metaphysics resting on the distinction between the sensible and the intelligible, consequently presumes the possible knowledge of super-sensible Reality. We may go back to Parmenides with his distinction of truth and appearance, to Plato's "divided line" in the *Republic* with its distinction between the lower visible and higher intelligible Reality, or to Aristotle's claim that metaphysics as "…the

---

[6] From a 2011 interview with Google cited here: http://www.telegraph.co.uk/technology/google/8520033/Stephen-Hawking-tells-Google-philosophy-is-dead.html (accessed June 15, 2016).

primary science treats of things which are both separable and immutable."[7] "Separable" here means "separable from matter." Hence metaphysics is a philosophical science by which the mind lays hold of objects which transcend the world of sense and change.

Accordingly, the modern anti-metaphysical project aims to show such a science is impossible. The main recourse is to effectively reduce knowledge to the empirical. The first phase of this project – presaged in some ways by Bacon and even Ockham -is naturally bound up with the British empiricist school of which the most radical and possibly formidable member was the eighteenth century Scotsman David Hume. The first essential step in his argument is his claim that sense experience is the exclusive sources of all our ideas and thus the empirical sets absolute limits to the possibilities of human thought:

> But though our thought seems to possess this unbounded liberty, we shall find, upon a nearer examination, that it is really confined within very narrow limits, and all this creative power of the mind amounts to no more than the faculty of compounding, transposing, augmenting, or diminishing the materials afforded us by the senses and experience. When we think of a golden mountain, we only join two consistent ideas, *gold* and *mountain*, with which we were formerly acquainted.[8]

The claims about the origin of ideas in sensation is a direct contradiction of epistemological rationalism with its notion of innate ideas, as other systems (Platonism, Augustinianism) which posit a non-empirical source of ideas. Taken in itself however this claim involves no necessary refutation of classical metaphysics per se. Indeed, Aristotle himself makes a similar claim about the origin of knowledge in the senses[9] which was elegantly summarized by St. Thomas Aquinas with his maxim *nihil est in intellectu quod non sit prius in sensu*[10] ("nothing is in the intellect which was not first in the senses"). For although Aristotle and his peripatetic and scholastic followers acknowledged the *origin* of ideas in the senses, they argue for the possibility of necessary inferences from the empirical to the metaphysical (e.g. God, substance, and the categories). This however is precisely the avenue to metaphysics that Hume will aim to cut off.

In general, nearly all the traditional arguments which infer the existence of a metaphysical object from sensible reality proceed by way of inference from causation (Aristotle's cosmological argument for God's existence being among the clearest examples.) The radicalism of Hume's empiricism is such that it puts the very idea of causation in question. Causation properly speaking embraces for Hume the idea of *necessary* connection – where the cause appears the effect follows of necessity; and where the cause is absent so is the effect. But says Hume this idea of necessity is not contained in the contents of sense experience:

---

[7] Aristotle. *Metaphysics*. E (1026a).
[8] David Hume. *An Enquiry Concerning Human Understanding*. (Mineola, N.Y.: Dover 2004): 10 (Section II).
[9] Aristotle. *On the Soul*. III.8.
[10] St. Thomas Aquinas. *De Veritate* II.3 in http://www.corpusthomisticum.org/qdv02.html. (accessed 10/1/2016).

11.1 David Hume's Skeptical Empiricism

> When we look about us toward external objects, and consider the operation of causes, we are never able in a single instance, to discover any power or necessary connexion; any quality, which binds the effect to the cause, and renders the one an infallible consequence of the other. We only find, that the one does actually, in fact, follow the other. The impulse of one billiard-ball is attended with motion in the second.[11]

In other words, though no true causation is contained in the empirical evidence of the senses as "...external objects as they appear to the senses, give us no idea of power or necessary connexion...".[12]

Necessary connection arising from the confusion of conjunction with causation not anything given in experience – "one event follows another, but we never observe any tie between them. They seem *conjoined* but never *connected*."[13]

The custom of observing a succession of events with a certain regularity gives rise to this *conjecture* of causal relation which is not genuinely justified by the evidence -" ...this idea of necessary connexion among events arises from a number of similar instances which occur of the constant conjunction of these events...".[14]

Hume's radical skepticism of causality cuts only major route from empirical to intelligible objects and so places the possibility of metaphysics in grave question. Hume draws out the implications of this in his discussion of other objects of traditional metaphysics. The Aristotelian concept of *substance* he regards as incoherent since all that is given in the senses are a collection of sense impressions not a genuine underlying unified subject:

> The idea of a substance as well as that of a mode, is nothing but a collection of simple ideas, that are united by the imagination, and have a particular name assign'd to them...[15]

Hume's critique of substance has implications even for our notion of a unified self. Is the Self a real substance underlying shifting experiences that provide a stable foundation for personal identity? Or is it merely a complex or as is often described a "bundle" of different and constantly shifting perceptions, emotions, thoughts, and experiences? Says Hume:

> This question we might easily decide, if we recollect what has already been prov'd at large, that the understanding never observes any real connexion among objects, and that even the union of cause and effect, when strictly exam'd, resolves itself into the customary union of ideas. From thence it evidently follows, that identity is nothing really belonging to these different perceptions...[16]

Hume's radical empiricism, and in particular his critique of causality thus portends the undermining of any "bridge" between the sensible world of empirical reality, and the traditional objects of speculative metaphysics – as for instance God, substance, and even the reality of a unified Self or soul.

---

[11] *An Enquiry Concerning Human Understanding.*(Mineola, N.Y.: Dover 2004):39 (Section VII).
[12] Ibid. 40.
[13] Ibid. 47.
[14] Idem.
[15] David Hume.*A Treatise of Human Nature* (Mineola, N.Y.: Dover, 2003 – Part I, Section VI).
[16] Ibid. Part IV, VI (p. 185).

11.2 Kant's Critical Philosophy

If we consider the project of Enlightenment philosophy in terms of the dethronement of metaphysics and the restriction of reason to the empirical realm, Hume's philosophy may be taken as the extreme point in this development. However, Hume's conclusions were so radical they put empirical science itself in jeopardy by questioning causality which is part of the necessary basis of the science. From the standpoint of the Baconian project with which the Enlightenment was linked this was problematic. While metaphysics is more or less irrelevant to the modern technological project, empirical science is absolutely vital. Hence the problem arises: how can traditional metaphysics as a speculative science of supersensible reality be overthrown without at the same time undermining the basis of empirical science?

Though he did not frame the issue in quite this way, the solution to this dilemma was central to the achievement of the most brilliant philosopher of the Enlightenment, Immanuel Kant. Kant had famously noted the striking personal effect of Hume's empiricist critique of metaphysics on the trajectory of his thought by saying that it was Hume who had "…first interrupted my dogmatic slumber…"[17] Hume had apparently demonstrated that reason cannot show any universal and necessary relationship between any presumed "cause" and "effect", as the conditions requisite to establish a causal relation – universality and necessity – are not given in sense experience.[18] This would seem to negate the concept of causality altogether.

Kant's radical solution to safeguard causality and other central concepts in empirical knowledge was to argue that universality and necessity are contributed by the knower and not by the things known:

> …the understanding does not derive its laws (a priori) from, but prescribes them to, nature…laws which we discover in objects of sensuous intuition (especially when these laws are cognized as necessary) are already held by us to have been placed there by the understanding…[19]

In other words, properties such as causality are universally valid with respect to nature because they are not received from nature but contributed to it as necessary conditions by the *human* faculty he calls "understanding." Thus, the Aristotelian categories such as substance and accident (in its varieties) and causality, which for the Greeks were *ontological* for Kant become *epistemological*. They are not attributes of being considered in itself, but rather they are forms of thought contributed by the mind which make human knowledge possible (albeit without themselves constituting knowledge.)

---

[17] Immanuel Kant. *Prolegomena to Any Future Metaphysics* (Ellington revision of Carus translation, Hacket Publishing, 1987): 5 (preface) . For a helpful overview of Kant's critique of metaphysics cf. Grier, Michelle, "Kant's Critique of Metaphysics", *The Stanford Encyclopedia of Philosophy* (Summer 2012 Edition), Edward N. Zalta (ed.), URL = <http://plato.stanford.edu/archives/sum2012/entries/kant-metaphysics/>. (accessed 10/26/2016).

[18] Ibid.3.

[19] Immanuel Kant. *Ibid.*: 62 (sections 36–37) – italics in original.

11.2 Kant's Critical Philosophy

That the categories for Kant are *a priori* and contributed by the understanding might seem to indicate kinship between critical philosophy and rationalist claims (e.g. Descartes) concerning innate ideas. Yet on the point crucial to the possibility of metaphysics, Kant stands squarely with Hume and the radical empiricist tradition in rejecting the possibility of supersensible knowledge. Every object of knowledge must for Kant possess not only a concept but also have a corresponding "empirical intuition" i.e. a direct sense experiences.

> To think an object and to cognize an object are thus not the same. For two components belong to cognition: first, the concept, through which an object is thought at all (the category), and second, the intuition through which it is given; for if an intuition corresponding to the concept could not be given at all, then it would be a thought as far as its form is concerned, but without any object, and by its means no cognition of anything at all would be possible… all intuition that is possible for us is sensible…[20]

Definitionally, on Kant's premise no metaphysical (i.e. supersensible) being can be an object of possible knowledge. But even more radically even "empirical intuitions" do not apprehend things themselves. The reason is that the manifold by which we perceive the sensory world -space and time – do not derive from experience but are rather the universal and necessary conditions also contributed by the knowing subject which make empirical reality possible. As Kant explains:

> Space is nothing other than merely the form of all appearances of outer sense, i.e., the subjective condition of sensibility under which alone outer intuition is possible…We can speak accordingly of space, extended beings, and so on only from the human standpoint. If we depart from the subjective condition under which alone we acquire outer intuition, namely the through which we are affected by objects, then the representation of space signifies nothing at all.[21]

Likewise, Kant argues for time as a subjective condition and form of representation contributed by the subject to experience rather than an objective property of things themselves:

> Time is not something that would subsist for itself or attach to things as an objective determination, and thus remain if one abstracted from all subjective conditions…Time is nothing other than the form of inner sense, i.e., of the intuition of our self and inner state. For time cannot be a determination of outer appearances; it belongs neither to a shape or position, etc., but on the contrary determines the relation of representations in our inner state.[22]

The reason is that the universal and necessary forms for the possibility of sense experience – space and time – are like the categories prescribed to nature by the knowing subject rather than received from it.

Now we are in a position to understand the core of Kant's case against the possibility of metaphysics. Metaphysics claims to be a speculative science which claims

---

[20] Immanuel Kant. *Critique of Pure Reason*. Translated by Paul Guyer and Allen W. Wood. (New York: Cambridge University Press, 2007): 254 (Doctrine of Elements, B146).

[21] Ibid.:177 (Transcendental Aesthetic, On Space, A26/B42-A27-B43).

[22] Ibid.:163 (Transcendental Aesthetic, On Time A33-B50).

a knowledge of realities which transcend the senses. But in Kant's schema concepts can only have applicability for demonstrative knowledge *within* the realm of sense experience. To apply concepts beyond the realm of the senses is what he calls "transcendental illusion."

> The principles of pure understanding we presented above should be only of empirical and not of transcendental use, i.e. of a use that reaches out beyond the boundaries of experience.[23]

This general idea provides the essential foundation for Kant's rejection as "illusion "the demonstrable knowledge claimed by special metaphysics (*metaphysica specialis*) for demonstrable knowledge of supersensible objects. Special metaphysics was a prevalent categorization rooted in Suarez and culminating in the continental rationalism of Leibniz and Wolff. Special metaphysics distinguished itself from ontology or general metaphysics (*metaphysica generalis*) in that while general metaphysics studies being in general (the being qua being of Aristotle or *esse commune* of the scholastics), special metaphysics has specific objects – the soul (rational psychology), the world (cosmology), and God (rational theology). For Kant each of these infers the existence of their objects by utilizing concepts of the understanding which only have applicability in the empirical sphere in a "transcendental "way – i.e. as if they can refer to super-sensible reality. Thus, for instance the cosmological arguments of Aristotle, Aquinas, and Leibniz inferred the existence of God from sensible reality arguing variously from the causal order to a First Cause or from the reality of contingent Beings to the reality of a Necessary Being. The error here according to Kant is applying notions like causality and contingency – valid in the realm of empirical science – outside the realm of possible experience.

> …in this cosmological argument an entire nest of dialectical assumptions is hidden, which transcendental criticism can easily discover and destroy…The transcendental principle of inferring from the contingent to a cause, which has significance only in the world but which outside it does not even have a sense…the principle of causality has no significance at all and no mark of its use except in the world of sense ; here however it is supposed to serve precisely to get beyond the world of sense…[24]

While such errors according to Kant are more or less unavoidable since it is part of the natural trajectory of reason of to arrive at the "unconditioned" – the ideas of God, the soul and the cosmos (hence he calls them "regulative ideas" since they guide or regulate the activity of reason), critical philosophy discloses such a "transcendental use of reason" as illegitimate at least with respect to demonstration. All that can be known of the world beyond the senses – things in themselves – is *that* they exist. Of reality beyond the world of sense and change charted by empirical science nothing at all can be known.

---

[23] Ibid. 386 /Transcendental Dialectic, Transcendental Illusion. (B353).

[24] Ibid, 572 (A609/B637--A610/B638 – Transcendental Dialectic- Impossibility of a cosmological proof.).

11.3 A Critique of the Critique: Kant's Anti-Metaphysical Revolution and Its Tension with Post-Newtonian Physics

It is indeed fitting that Kant opens his *Critique of Pure Reason* with a quotation from Sir. Francis Bacon. With Kant, the Baconian revolution may be said to have reached its culmination. Bacon begins the turn away from traditional metaphysics on the grounds of its practical fruitlessness, while for Kant metaphysics (as traditionally understood) is not merely fruitless but also impossible. And while Bacon was central in the establishment of the method of empirical science and in privileging its role, Kant limits theoretical reason to the empirical sphere restricting metaphysics understood as a science of super-sensible reality, even while providing a philosophical grounding for empirical science.

To be sure there are necessary qualifications to this portrayal of Kant as an enemy of metaphysics. There is a sense in which Kant with his apriorism claimed to *reconceive* metaphysics as a study of the *conditions* of knowledge rather than as a theoretical science of immaterial reality. But this is more or less to overthrow its original meaning as defined by Aristotle. It is also true that for Kant metaphysical ideas reappear as *practical* postulates – hence for example the ideas of God and freedom are is requisite to ground moral life. Yet as Leo Strauss notes this very shift from theoretical to practical reason is highly significant. We may say it is one symptom of the modern devaluation of the theoretic life. For Aristotle as we saw the theoretical was privileged over the practical, and metaphysics seen as the culmination and highest activity of the theoretical life. Kant here privileges practical reason as the place where metaphysical ideas like God and freedom receive their relevance.

Did Kant succeed? For all the brilliance of its execution there are serious problems to the Kantian critique of metaphysics from both the philosophical and (more ironically) scientific perspectives. Kant argues that that the thing in itself is a "true correlate"[25] of the phenomena which appear to our sensibility. This implies that the "thing as it appears" requires an external cause lest, says Kant, we would have the "…absurd proposition that there is appearance without anything that appears."[26] Yet according to Kant causality itself is a "pure category of understanding" without legitimate use beyond the empirical realm. The thing in itself and the restriction of the categories to sense experience seem to be in contradiction.

This problem has long been recognized with Kant's immediate successors Historians of philosophy will consider Schulze's *Aenedismus*, Fichte, etc... In this regard W.T. Stace notes that:

> Everyone saw at once that the thing -in -itself… is a flat self-contradiction. Its existence is assumed because Kant assumed there must be an external *cause* of our sensations. On the one hand, therefore, the thing-in -itself is alleged to be the cause of appearances. On the

---

[25] Ibid.162 (Transcendental Aesthetic, On Time B45/A30).

[26] Ibid.Bxxvi-vii (Preface).

other hand, however, it cannot be a cause, because cause is a category of our minds, and the categories do not apply to the thing in itself.[27]

To resolve this contradiction there are only two paths – to affirm or reject the thing in itself. But whichever horn of dilemma one chooses brings us back to the inescapable questions of metaphysics. If we affirm the thing in itself it seems one must drop the requirement that the categories have only an empirical use, since we infer that a super-sensible being (the thing in itself) exists on the basis of inference from sensible beings. And if it is indeed is possible or necessary to infer the existence of one super-sensible reality how can one then dogmatically close off the possibilities of other avenues by which the existence of super-sensible objects is inferred from the evidence of the senses (e.g. the inference of a necessary being from contingent beings)? This puts in question Kant's entire critique of metaphysics.

But what if one gives up the thing in itself? This apparent solution raises another specter -the *reductio ad absurdum* of solipsism. For if there is nothing that can be affirmed to exist independently of the mind then how does one acquire certitude that that the mind does not simply constitute all reality? The solution of the German idealists (Fichte, Schelling, Hegel) was to affirm that in one sense Mind DOES constitute reality, but that Mind is not restricted to the individual subjectivity. Rather they sought to overcome the division of subject and object (seeing them both as moments in the unity of a higher, Absolute Mind. This overcomes the distinction between the thing as it appears and the thing in itself. But this is merely a route to another form of metaphysics – absolute idealism.

Perhaps ironically however, the gravest problems for the Kantian project may have arisen not from the metaphysical objections which may be raised to it, but from the forward progress of empirical science. The Kantian critique of metaphysics aimed as was mentioned at restricting the legitimate domain of reason to the empirical realm, i.e. to empirical science. Yet by "hitching his wagon" to a *particular* empirical scientific model – the Newtonian world system – Kant's opened the possibility that his philosophical contentions could be brought into question and undermined by the forward progress of empirical science. Had Kant merely accepted the Newtonian conceptions of space and time as absolute, Euclidian, and independent of objects and events as *empirical hypotheses* it might have posed no particular problem for his philosophical system. But by *canonizing* Newtonian conceptions of space and time as universal, necessary, and valid *a priori*, Kant's position becomes vulnerable to empirical refutation as the Newtonian world system became progressively antiquated.

Among the Newtonian principles that Kant endeavored to render as *a priori* and universal are first that space is absolute and as such independent of matter (i.e. "empirical intuitions"); secondly that time is absolute and independent of the empirical realities which inhabit it; third that space is Euclidian and that the laws of

---

[27] W.T. Stace. *The Philosophy of Hegel* at https://archive.org/stream/W.T.StaceThePhilosophy OfHegelDoverPress1955/W.T.%20Stace-The%20Philosophy%20of%20Hegel-Dover%20 Press%20(1955)_djvu.txt (accessed November 4, 2016).

11.3 A Critique of the Critique: Kant's Anti-Metaphysical Revolution and Its Tension…

Euclidian geometry are necessary modes of representing spatial reality; and fourth that space and time are independent not only of matter but of each other. Yet subsequent developments in modern physics, have cast doubt on each and every one of these elements in the Newtonian world system which Kant had attempted to canonize as universal, necessary, and a priori.

Kant's idea that space and time are simply a kind of "theatre" existing independently of the things and events which occur in them is derived directly from the Newtonian world system. In the Cartesian system space can be thought only in terms of "relations to other bodies"[28] and cannot be thought to exist as something distinct for corporeal substance.[29] Newton by contrast argues for the existence of an absolute space and time. In the Scholium of his *Principia* Newton states that:

> Absolute, true, and mathematical time of itself, and from its own nature, flows equably without relation to anything external.

And:

> Absolute space, in its own nature, without relation to anything external, remains always similar and immovable.[30]

This Newtonian separability of space and time from empirical things and events provides the conceptual foundation for Kant's distinction between space and time as "pure forms" distinct from and independent of the "empirical intuitions." As Kant put it:

> One can never represent that there is no space, though one can very well think that there are no objects to be encountered in it. It is therefore to be regarded as the condition of the possibility of appearances, not as a determination dependent on them, and is an *a priori* representation that necessarily grounds outer appearances.[31]

Majors cracks in this Newtonian edifice had begun to become evident by the nineteenth century. Perhaps the most important physicist since Newton, James Clerk Maxwell achieved an essential breakthrough by providing a set of equations to account for electromagnetic interactions, – electricity and magnetism now being correctly understood as two elements of the same phenomena. Electromagnetic waves (including) light however were found not to behave in the way that the classical rules of Galilean relativity had predicted. According to classical mechanics while space and time are absolute, velocity is relative to the motion of the observers (e.g. an object going on at 100 mph relative to one stationary to it ought to be going only 50 mph relative to one moving at 50 mph in the same direction.) This principle however did not seem to function with respect to the propagation of light as was

---

[28] Descartes. *Principles of Philosophy*, 2.13 in http://www.earlymoderntexts.com/assets/pdfs/descartes1644part2.pdf (Nov. 13, 2017).

[29] Ibid. 2.11.

[30] Newton . *Principia*. From the 1729 translation by Andrew Motte https://plato.stanford.edu/entries/newton-stm/scholium.html (November 13, 2017).

[31] Kant. *Critique of Pure Reason (supra)*, A24/B39, p. 158.

confirmed when the Michelson-Morely experiment in 1887 indicated that the speed of light is invariant as measured from all reference frames.

Physics was thus left with a contradiction between the classical principles which implied the measured velocity of anything is contingent on the relative motion of observers with respect to it, and the empirically observed fact that the speed of light (in a vacuum) is apparently an absolutely fixed constant to all observers.[32] It was left to Albert Einstein to resolve these apparent contradictions through the special and general theories of relativity. The cost however was precisely giving up the idea of space and time as absolute and independent from matter and motion (Kant's system treats the latter as "empirical intuitions" in contrast to the *a priori* forms of space and time.)

At the point in which relativistic effects become manifest (i.e. as objects approach the speed of light), the special theory of relativity shows that length contracts – showing the that a spatial dimension is contingent on the motion of matter. The general theory of relativity moreover treats gravity as the curvature of space by mass correlated to the mass of the objects which inhabit space, showing that so far from being a manifold independent of matter, space is integrally affected by it. Einstein himself put the matter succinctly:

> According to the general theory of relativity, the geometrical properties of space are not independent, but they are determined by matter.[33]

The same is equally true of time. While Newtonian physics is a reasonable approximation of our universe at the everyday common-sense level, at the point at which relativistic effects become relevant (as objects move closer to the speed of light), any conception of absolute space and time must break down due to time dilation (the slowing down of time as objects approach the speed of light. This means both that different objects depending on their velocity will each have their own time. "simultaneity" which is a coherent concept if time were absolute, now becomes a relative concept. The relativity of simultaneity means that events a and b may be perceived as simultaneous in one frame of reference but as following each other in a different one. As Einstein put the matter:

> Every reference body (co-ordinate system) has its own particular time; unless we are told the reference-body to which the statement of time refers, there is no meaning in a statement of the time of an event.[34]

Kant likewise argues that three-dimensional Euclidian geometry is a universally necessary and *a priori* mode of representing reality:

> Geometry is a science which determines the properties of space synthetically and yet *a priori*...geometrical propositions are all apodictic i.e., combined with consciousness of

---

[32] Max Born. *Einstein's Theory of Relativity*.(New York: Dover, 1962): 225ff.

[33] Albert Einstein. *Relativity: The Special and General Theory* "The Structure of Space Time According to the General Theory of Relativity", 32 http://www.bartleby.com/173/32.html (Accessed 4/24/18).

[34] Albert Einstein. *Relativity.* Chapter 9 "The Relativity of Simultaneity" 7 http://www.bartleby.com/173/9.html (Accessed April 23, 2018).

11.3 A Critique of the Critique: Kant's Anti-Metaphysical Revolution and Its Tension…

their necessity e.g., space has only three dimensions; but such judgements cannot be empirical nor judgements of experience, nor inferred from them…[35]

However, to explain the empirical phenomena like the effect gravity, Einstein was forced to employ non-Euclidian (Riemannian)geometry. This is at the least a challenge to the idea that there is an *a priori* universality and necessity in Euclidian space, for if this were the case it is strange that for contemporary physics that as Einstein put it "…the space-time continuum cannot be regarded as a Euclidian one…"[36] Following Hermann Minkowski, contemporary physics moreover treats space-time moreover as (at least) a *four-dimensional* continuum. Besides challenging the necessity of Euclidian space, this challenges another central pillar of the Newtonian-Kantian model – the independence of space and time from each other. As absolute and independent space and time have receded, the idea of an integrated space-time continuum has replaced it. For while the time of an event cannot be posited in a way which is true and invariant for all reference frames, the space-time interval, a mathematical variable which relates space to time can be. Minkowski deriving his ideas from these consequences of relativity showed that time can be thought of as a dimension within the geometry of a unified space-time continuum. As he famously put it:

> From now on space-in-itself and time-in-itself are destined to be reduced to shadows, and only a sort of union of the two will retain an independent existence.[37]

All of this at the least raises formidable problems for the Kantian conception. How can a *particular* conception of space and time (in which space and time are absolute and distinct from each other, space is Euclidian, and both space and time are unaffected by matter) be the *universal* and *necessary* mode for the empirical representation of reality, if this conception fails to actually correspond to the observed empirical realities? Thus, crucial premises of Kant's critique of metaphysics have thus been rendered problematic by the modern crisis of the Newtonian world system in physics. But if the premises that underlie Kant's critique of metaphysics are incorrect, it may well be premature to assume this critique wrote the obituary of Western metaphysics.

The great physicists who brought down the Newtonian world system – among others Max Planck, Albert Einstein, Werner Heisenberg, and Niels Bohr – were really "natural philosophers" in the older language, with more theoretical casts of minds then inventors and technologists. Nonetheless, the valorization of science in the modern age is almost invariably tied up with an ultimate concern for application. Einstein himself may have been impelled by theoretic wonder at the workings of the universe. However, it is the Baconian applicability of his principle of mass-energy

---

[35] Kant. *Critique of Pure Reason (supra)*, B41, 176.

[36] Einstein. *Relativity*. 27. http://www.bartleby.com/173/27.html (accessed 4/24/2018).

[37] Hermann Minkowski. "Space and Time" a published lecture in *The Monist* Volume XXVIII (1918) in https://archive.org/stream/monistquart28hegeuoft/monistquart28hegeuoft_djvu.txt (accessed April 23, 2018).

equivalence for harnessing the nuclear power of the atom for fuel and weaponry on which the fame of his theory also rests.

Despite unresolved difficulties in the Kantian project from the sides of both philosophy and science, the historical impact of this most brilliant philosopher of the Enlightenment is beyond all question. The whole pre-Kantian tradition of Western metaphysics could now be comfortably consigned to the category of the "pre-critical" or "dogmatic" phase of philosophy. While the particular sciences seemed to advance by leaps and bounds, philosophy -which is to say the universal science -seemed to stumble, fundamentally doubting its own possibilities. In Kant, the Baconian revolution comes to completion. The primacy of theoretical reason and metaphysics gives way to that of practical reason and empirical science. The Queen of the Sciences has been indeed cast from her throne.

References

Aquinas, St. Thomas. 1961. Trans. John P. Rowan. Anthony Kenny. Ed. http://dhspriory.org/thomas/Metaphysics.htm. Accessed 16 June 2016.

———. 1995. *Commentary on Aristotle's Metaphysics.* Trans. John P. Rowan. Notre Dame, Indiana: Dumb Ox Books.

———. De Veritate at University of Navarra. http://www.corpusthomisticum.org/qdv02.html. Accessed 1 Oct 2016.

Aristotle. *Metaphysics* 1933 (2003 reprint). Trans. Hugh Tredennick. Loeb Classical Library, Harvard University Press.

Benedict XVI. 2006. Benedict XVI. "Faith, Reason, and the University: Memories and Reflections" often called "The Regensburg Lecture" At the Vatican. http://w2.vatican.va/content/benedict-xvi/en/speeches/2006/september/documents/hf_ben-xvi_spe_20060912_university-regensburg.html

Descartes, Rene. 1644. Principles of Philosophy, http://www.earlymoderntexts.com/assets/pdfs/descartes1644part2.pdf. 13 Nov 2017.

Einstein, Albert. 1920. *Relativity.* The Relativity of Simultaneity. Translation Robert W. Lawson. New York: Henry Holt. http://www.bartleby.com/173/9.html. At Bartleby.com (2000) Accessed 23 Apr 2018.

Grier, Michelle. 2012. "Kant's Critique of Metaphysics." At The Stanford Encyclopedia of Philosophy. (Summer 2012 ed.) Edward N. Zalta (ed.) http://plato.stanford.edu/archives/sum2012/entries/kant-metaphysics. Accessed 26 Oct 2016.

Hegel, G.W.F. 2001 *The Philosophy of History.* Trans. J. Sibree. Kitchener, Ontario: Batoche Books. http://socserv2.socsci.mcmaster.ca/~econ/ugcm/3ll3/hegel/history.pdf. Accessed 15 Jun 2016.

Hume, David. 2003. *A Treatise of Human Nature.* Mineola: Dover.

———. 2004. *An Enquiry Concerning Human Understanding.* Mineola: Dover.

Kant, Immanuel. 1987. Prolegomena to Any Future Metaphysics. James Ellington revision of Paul Carus translation, Indianapolis, Hackett Publishing.

———. 2007. *Critique of Pure Reason.* Translated by Paul Guyer and Allen W. Wood. New York: Cambridge University Press.

Minkowski, Hermann.1918. *Space and Time* a published lecture in *The Monist* Volume XXVIII at https://archive.org/stream/monistquart28hegeuoft/monistquart28hegeuoft_djvu.txt. Accessed 23 Apr 2018.

References

Newton, Sir. Isaac. 1729. *Principia* (Scholium) Trans. by Andrew Motte. Stanford Encyclopedia of Philosophy. https://plato.stanford.edu/entries/newton-stm/scholium.html. Accessed 13 Nov 2017.

Stace, W.T. 1955. The Philosophy of Hegel. New York. Dover. At Archive.org. https://archive.org/stream/W.T.StaceThePhilosophyOfHegelDoverPress1955/W.T.%20Stace-The%20Philosophy%20of%20Hegel-Dover%20Press%20(1955)_djvu.txt. Accessed 4 Nov 2016.

Warman, Matt. May 17, 2011. Stephen Hawking Tells Google 'philosophy is dead'. *The Telegraph*. http://www.telegraph.co.uk/technology/google/8520033/Stephen. Accessed May 2018.

Chapter 12
Progressivism, Commerce, and the Triumph of Machine Civilization

"Background with binary code and face. Background of technology and hacker." –Carlos Castilla.–www.shutterstock.com

12.1 The Idea of Progress

"Theoretical man" has found himself increasing homeless in the modern world. The Greek theoretical ideal was challenged during the Enlightenment not only by the anti-metaphysical critiques we have discussed, but even more by the vast social revolution it spawned. Three elements of the modern world view merit special examination in this connection – the idea of progress, the new valorization of commerce and technology, and cultural egalitarianism.

Nowhere was the influence of Bacon on the Enlightenment so deep and revolutionary as in contributing a new conception of time and history. Prior to Bacon and the Enlightenment, European civilization (like all others) for inspiration to real or imagined golden ages of the past, rather than in the future. In this respect it is interesting to compare the Enlightenment with the earlier kindred movement of the Renaissance. The Enlightenment shared with the Renaissance a keen conviction regarding the dignity of man and an optimistic confidence in human powers, its humane and tolerant spirit, and a concern for naturalism in both art and science. This is doubtless is why Berdyaev among others correctly saw in it a later phase of the same European humanism. Yet the Renaissance culture deferred to Greek and Roman antiquity and looked almost piously to the wisdom of the ancient sages of the classical world – an attitude of reverence given visible form in Raphael's famous *School of Athens*. Indeed, as the name implies the Renaissance was a "rebirth" of the classical intellectual and aesthetic ideals. A measure of the vast paradigm shift which had occurred in attitudes toward the past – including Greco-Roman antiquity – in the intervening centuries is shown in Voltaire's *Philosophical Dictionary* under the entry "Precis of Ancient Philosophy":

> I have consulted all the adepts of antiquity, Epicurus and Augustine, Plato and Malebranch and I have remained in my poverty. Maybe in all the philosophers 'crucibles there are one or two ounces of gold; but all the rest is residue, dull mud, from which nothing can be born.[1]

In this lay the difference. While the Renaissance looked to the ancient past as a Golden Age of learning before the world relapsed into barbarism with fall of the Roman Empire, the Enlightenment *philosophes* regarded *their own time* as superior to all that came before, and believed the golden age lay ahead in futurity. As the Marquis de Condorcet writes

> Everything tells us we are approaching the era of one of the grand revolutions of the human race…the present state of knowledge assures us it will be happy.[2]

History was thus entirely reconceived. But why and whence did the radical idea of progress emerge? Reinhold Niebuhr argues that in spite of the avowedly secular character of the Enlightenment, Christianity was a central – if subterranean – influence:

> The idea of progress is possible only upon the ground of a Christian culture. It is a secularized version of Biblical apocalypse and of the Hebraic sense of a meaningful history… But since the Christian doctrine of the sinfulness of man is eliminated, a complicating factor in

---

[1] Voltaire. *Philosophical Dictionary*. https://history.hanover.edu/texts/voltaire/volpreci.html (Accessed January 4, 2017).

[2] Marquis de Condorcet. *Sketch for a Historical Picture of the Human Mind* in *The Portable Enlightenment Reader* Isaak Kramnick(ed.) (Penguin U.S.A., 1995): 394–395 Hereafter "Kramnick".

the Christian philosophy is removed and the way is open for simple interpretations of history ...which fail to do justice either to the unique freedom of man or to the daemonic misuse which he may make of that freedom.[3]

The Greek theoretical mind found the locus of meaning in the universal and the eternal and found little in the historical process beyond particularity and contingency.[4] The idea of a linear history unfolding toward a higher purpose and greater justice belongs to what Christianity inherited from Israel, and what the Enlightenment inherited from Christianity.

Judeo-Christianity's linear conception of time is thus a precondition for the idea of progress. But what gave force to the idea of progress was more than anything else scientific and technological advancement which the men of the Enlightenment could see before their eyes – the fulfillment of Bacon's prophecy of concerning technological progress made possible through scientific advancement. The steady growth in human power over nature made possible through ever more efficacious machines seemed to give eloquent testimony both to human intelligence and to optimistic hopes concerning the improvement of the human condition.

The *philosophes* had as yet no reason to suppose that moral progress would lag behind material progress, nor did they yet see – as the twentieth century would – the possibilities for a "daemonic misuse" of technological power. As the Marquis de Condorcet writes:

> The sole foundation for the natural sciences is this idea, that the general laws directing the phenomena of the universe...are necessary and constant. Why should this principle be any less true for the development of the intellectual and moral faculties of man than for other operations of nature?[5]

Improved knowledge and the advancement of science would indeed lead to "... the true perfection of mankind".[6]

While this view of progress included, indeed presupposed the advancement of knowledge, one can see also why the Greek theoretical ideal also came to be seen as something antiquated as a source of wisdom or inspiration. If the golden age lies in the future, why look to the wisdom of the ancients for guidance? Moreover, the *type* of knowledge Voltaire and Condorcet saw as progressive was principally of the Baconian variety – scientific-technological changes that would lead to advances of practical utility for mankind. The basically utilitarian understanding of knowledge, so different from the classics was equally influenced by the rise of the commercial society, which in turn became the principal engine for technological development.

---

[3] Reinhold Niebuhr. *The Nature and Destiny of Man*. Volume I (Charles Scribner, 1964): 24.
[4] Cf. Aristotle. *Poetics* 1451b5-7.
[5] Marquis de Condorcet. Kramnick, 26 Hereafter "Kramnick".
[6] Ibid, 27.

12.2 The Valorization of Commerce and the Mechanical Arts

We have already seen that the classical philosophers distinguished among goods, elevating the noble and excellent above the merely useful and necessary. The noble is superior because it is an end in itself, while what is useful is good merely as a means to some good beyond itself. This perspective naturally influenced the ancient evaluation of commercial activity which aims at money making. While many *treat* money as if it were a supreme good, Aristotle forcefully rejects the "life of money making" which would make of lucre the supreme good. Money as an instrument of use has no intrinsic value beyond its utility:

> The life of money making is a constrained kind of life, and clearly wealth is not the Good we are in search of, for it is only a good as being useful, a means to something else.[7]

This classical Greek concept entered into aristocratic Roman ideals which saw commerce (as well as manual labor) as well not proper to the Roman gentleman, who being free to pursue more noble activities (such as the liberal arts), should not occupy his with lower activities. We will recall the view of Seneca:

> …I respect no study, and deem no study good, which results in money-making. Such studies are profit-bringing occupations, useful only in so far as they give the mind a preparation and do not engage it permanently. One should linger upon them only so long as the mind can occupy itself with nothing greater.[8]

In the Middle Ages, the classical-aristocratic attitudes joined with Christian ascetical attitudes toward acquisitive wealth-seeking to produce a rather negative evaluation of commerce. We may think of St. Thomas Aquinas's today striking statement that:

> Therefore business, considered in itself, has a certain turpitude, inasmuch as it does not concern in its own nature any honorable or necessary end.[9]

Aquinas's reasoning seems to follow logically from Aristotle's pre-suppositions. Considered *in itself* business is useful rather than noble and honorable because it is not done for its own sake. Hence it is a servile rather than liberal type of activity.

---

[7] Aristotle. *Nicomachean Ethics* I. 5.8.

[8] Lucius Annaeus Seneca, *Epistulae Morales*, LXXXVIII. English translation: Moral Letters to Lucillius:Letters from a Stoic. Aegitus. Can be accessed here: https://play.google.com/store/books/details/Seneca_Lucius_Annaeus_Moral_letters_to_Lucilius?id=6ykJAwAAQBAJ (Accessed October 25, 2018). I believe this is the Richard Mott Gumere translation.

[9] Ideo negotatio, secundum se considerata, quandam turpitudinem habet, inquantum non importat de sui ratione aliquid honest finem honestam vel necessarium. St. Thomas Aquinas. *Summa Theologiae*. ST II-II- 77 – the Latin text I accessed here: http://www.corpusthomisticum.org/sth3061.html#42226 (accessed 1/2/2016). The English translation is my own.

12.2 The Valorization of Commerce and the Mechanical Arts

Modern thinkers however completed inverted this evaluation. For a contrast with Seneca consider David Hume's argument for the moral as well as material benefits of commerce:

> And this perhaps is the chief advantage which arises from a commerce with strangers. It rouses men from their indolence; and presenting the gayer and more opulent part of the nation with objects of luxury, which they never dreamed of, raises in them a desire of a more splendid way of life than what their ancestors enjoyed. And at the same time, the few merchants who possess the secret of this importation and exportation, make great profits; and becoming rivals in wealth to the ancient nobility, tempt other adventurers to become their rivals in commerce.[10]

The late eighteenth century likewise saw the birth of modern free market economics with the publication of Adam Smith's *The Wealth of Nations* as well as the preliminary stages of the industrial revolution which would forge the mighty engine of modern technological capitalism. This revolution in thought is among the most remarkable. The rising position of both mechanical arts and of the profit seeking activities of commerce in social estimation seems to be mutually re-enforcing, as technology acts as a stimulus to commerce and the competitive mechanisms of the market act as a spur to rapid technological innovation. Improved transportation and communications technologies make commerce ever more efficient, and the prospect of profit achieved through new technologies became itself spur to further technological inventiveness. The success of the Scottish Enlightenment of Smith and Hume in valorizing the commercial society is thus a natural corollary of the Baconian revolution and its valorization of the mechanical arts.

This implied further a shift from the classical evaluation of *labor* as opposed to leisure. For the classics labor has no noble end in itself and its value such as it has is derived from its ability to produce leisure. The most blessed are the gentlemen who are "free" from servile labor entirely and can devote themselves to the liberal arts proper to free men. Note that Hume's apologetics for commerce rests on the idea that it encourages industry and makes men less indolent. Here especially in the Northern nations we have the idea of labor and the "Protestant work ethic" as the source of value with leisure receiving progressively less attention or even associated with sloth. Even the new value of the once generally condemned trait of *acquisitiveness* emerged. As reviled by the classical philosophers, as by medieval Catholic Christianity acquisitiveness tended to be associated with the vice of Greed (πλεονεξία) – the blind and vulgar "drone lusts" for wealth decried by Plato, or the *avaritia* denounced by the Latin scholastics. Hume however sees in the spirit of commerce and the competitive drive to acquire wealth something virtuous, a spur to industry and improvement.

Underlying all these elements are vast social changes. By the eighteenth century the commercial classes, the "bourgeoisie" were the rising force in European (and nascent American) society, while the landed aristocracy was already well in decline.

---

[10] David Hume. *Essays, Moral, Political, and Literary.* Part II Essay I "Of Commerce". http://oll.libertyfund.org/titles/hume-essays-moral-political-literary-lf-ed/simple#lf0059_label_371 (Accessed October 28, 2018)

This shift naturally destabilized the social and political order of the West leading to the democratic revolutions (discussed next.) Logically this shift toward a new political economy "supercharged" the modern values of pragmatism, technical efficiency, industry, work ethic, and utility, while the older classical ideals seemed to be gradually superseded. And here lies the essential problem from the classical perspective.

The classical world view resumed during the Renaissance saw the liberal arts, fine arts, and philosophy (in short high culture) as broadly representing "the noble". These activities seek goods excellent in themselves such as truth, virtue, and beauty. An activity such as commerce seeks merely "the useful" and "the profitable". So long as commercialism recognized its place in this classical hierarchy of values there is no inherent contradiction. Indeed, successful commerce conceived as a means and not as an end could be a boon for culture as one sees for instance in the vast artistic and literary patronage of the Medici – originally a banking family.

The issue arises from what British historian Christopher Dawson called "The Bourgeois mind".[11] Where the commercial form of evaluation becomes *dominant* "the useful" and "the profitable" come to be seen as goods in themselves. This necessarily had a corrosive effect on the European ideal of high culture. The goods attained by philosophy are not generally profitable or useful in the coarsely economic sense, and even if they were their value does not consist in this. The hegemony of "profitability" and "utility" as the *sole* criteria of value must necessarily yield a culture characterized by philistinism (contempt for culture) and materialism, and this is precisely why "bourgeois"(quite apart from the critique of the even more materialistic Marxists) acquired such a pejorative connotation in the nineteenth century. Kept within reasonable bounds the "bourgeois spirit" is a spur to innovation and dynamism; but where it is *hegemonic* it leads to a crassly economic conception of human existence which recognizes no valid human aim beyond mere material comforts, goods, and possessions. It has built a culture focused not only of utility, but of *entertainment*. Moreover, the utilitarian perspective of profit tends to have a levelling effect on culture since its logic is the maximizing of profits by pleasing the widest possible mass market.

12.3 Machine Civilization

Our age is without question the age of the machine. This is the ripest fruit of the Baconian dream in which technological power is the right hand of scientific inquiry and knowledge. Newtonian physics combined with practical machines like James Watt's stream engine to bring about the industrial revolution which replaced hand craftmanship with mechanized production in steam powered factories and the replacement of man and horse transport with steam powered ships and trains. The great theoretical advancements in the knowledge of electromagnetism associated

---

[11] Christopher Dawson. *Catholicism and the Bourgeois Mind.* https://www.catholicculture.org/culture/library/view.cfm?recnum=2580 (accessed Jan 7, 2017).

with the work of Faraday and Maxwell led rapidly to the development of new communication devices like the telephone and telegraph, the X-Ray, the radio, photography and eventually television. Increasing knowledge of the atom led to the development of the transistor and the microchip opening our digital age of calculators and computers, digital smart phones, and the internet. And of course atomic theory as with Einstein's conception of mass-energy equivalence led to the development of nuclear weapons and nuclear power.[12]

Progressively every domain of human life has become ever more mechanized and technologized. Communication, transportation, medicine, and production have all become machine processes. The human environment in terms of temperature and light is controlled by devices. War making has become an ever more impersonal and technical process often as with cruise missiles, high altitude bombers and drones fought between people in remote locations who never meet on the battlefield. Research and education and even social life take place ever more on the internet. Today we expect that even in the near future year by year new medicines will better treat diseases, computers will become more capable and powerful, transportation and communication faster and more efficient, national GDPs to grow, and our instruments and weapons ever more efficaciously destructive of nature and our fellow man. Such are the standard expectations within a Baconian civilization.

We have only begun to come to terms with the vast impact of the machine on human life. One of the most radical impacts of machine civilization is its effect on the very notion of time. Traditional agrarian culture was oriented around the recurrent patterns of nature. Logically it tended to produce static or cyclical models of time, and the expectation of relative permanence and stability. Human life particularly in its modern urban and technical setting is unlike traditional agrarianism in that it no longer operates according to the natural cycles of nature – the rising and falling of the sun or the changes of the seasons. Instead, human life is subject to the time of the artificial time of the digital clock in the laboratory, office, or factory. The machine produced a dynamic conception focused on constant instability, change, and innovation. The idea of progress is largely the consideration of human history from the vantage point of the machine. Computers today are far more efficacious than they were 50 years ago, and we expect than 50 years in the future they will be more advanced than today. From the continual advancements in the efficacy of the machine, we infer a directionality to human history toward improvement. But equally this dynamism which has so quickened the pace of human life introduces a radical inconstancy where the past is constantly erased, while the present quickly becomes obsolete and is replaced by the new. Already in the nineteenth century, Marx in spoke of this aspect which the new technologically driven processes of economic production had introduced into human life with their relentless drive for innovation:

> Constant revolutionizing of production, uninterrupted disturbance of all social conditions, everlasting uncertainty and agitation distinguish the bourgeois epoch from all earlier ones.

---

[12] James Rivington "7 Scientific Breakthroughts that unlocked the Age of Technology" August 23, https://www.techradar.com/news/world-of-tech/future-tech/physics-and-technology-1174316 (Accessed May, 2018).

> All fixed, fast frozen relations...are swept away, all new formed ones become antiquated before they can ossify. All that is solid melts into the air...[13]

Long before the digital revolution, Berdyaev put the matter as such:

> The overwhelming technical achievements of the nineteenth and twentieth centuries produced the greatest revolution in human history, far more important than all political revolutions. They brought about a radical change in the whole rhythm of human life, a break with the natural cosmic rhythm, and the appearance of a new mechanically determined rhythm. Machinery destroys the old wholeness and unity of human life, it tears away, as it were, the human spirit from organic flesh and mechanizes the material life of man.[14]

Like Heidegger if more ambivalently, Berdyaev was here concerned with the uprooting by technology of man from "telluric" life – i.e. the earth and nature.:

> Machinery acquires a universal significance, and puts its seal upon everything, making all into its own semblance. ...Life is no longer bound up with the earth, animals, and plants, and becomes connected with machinery, with a new reality that seems not to have been created by God.[15]

This theme has of course been forcefully taken up by modern environmentalism. Yet, in spite of these harsh statements, Berdyaev's attitude to technology was not wholly negative, so much as ambivalent. He blamed technology for much of the materialism and ugliness of modern life, yet recognized its testimony to human creativity, its role in battling the ancient scourges of poverty and illness, and the possibilities it had for bringing about greater freedom from drudgery into human life. If Heidegger's principal fear was that technology uproots man from the earth, Berdyaev saw this process as two sided. The paradigm shift from a telluric to a mechanized economy can be either dehumanizing or liberating:

> Technical progress has a two fold influence on man's social and moral life. On the one hand it lessens spirituality and makes life more material and mechanical. On the other hand it stands for dematerialization and disincarnation, opening up possibilities for greater freedom for the spirit.[16]

The difference between Heidegger and Berdyaev on this point may relate to Berdyaev's Christianity which affirms the supremacy of the human spirit over nature, as against the romantic elements in Heidegger with its pining for the earth and nature. The supine or worshipful attitude to nature is more characteristic of a quasi-pagan worship of the earth and nature, then of Christianity, even at its most critical toward uncontrolled technologism.

Berdyaev's ambivalence is understandable. Human life was once largely at the mercy of natural catastrophes such as the viruses and bacteria which have now been largely eradicated or drastically curtailed. Yet technological advancement has also threatened humanity with the specter of nuclear annihilation and made possible the modern totalitarian Marxist and Fascist regimes. Technology in the Baconian understanding is power, and power can be used for either constructive or destructive

---

[13] Ibid. 223.
[14] Berdyaev. *The Destiny of Man*.226.
[15] Ibid. 227.
[16] Idem.

ends. This makes it unclear whether the machine is a friend or foe to man. This question hinges on which humanity can place the machine under moralized human control.

Heidegger among modern philosophers who delved deeply into the problem of technology was skeptical. One of the reasons that he questioned the sufficiency of the definition of technology as instrumentality was precisely that instrumentality implies human mastery of technicity:

> Everyone knows two statements that answer our question [concerning technology]. One says: technology is a means to an end. The other says: technology is a human activity. The two definitions belong together. For to posit ends and procure and utilize the means to them is a human activity.[17]

Heidegger denies this saying that "Technicity in its essence is something that man does not master by his own power."[18]

If so we must admit that Heidegger is speaking of *the end of humanism*. The intellectual forefathers of modern technological civilization, above all Bacon and Descartes presumed that technology could remain precisely a benign instrument for the human domination of nature and the material uplift of the human condition. Heidegger is arguing that technology has become an all-consuming frenetic activity which no longer operates according to conscious human ends, and which controls and conditions man rather than reverse. Like the Sorcerer's apprentice it would seem that technology is an activity which contrary to all initial intentions has *slipped* entirely beyond human control. Berdyaev refers to this process of technology slipping out of human control and becoming malevolent as a form of dark magic:

> Technical developments are morally neutral up to a point. When they reach a certain level they lose this neutral character and may turn into black magic if the human spirit does not subordinate them to a higher purpose.[19]

The way in which the frenzy of technological and industrial development in the nineteenth century culminated in the destruction of two World Wars, and the looming threat of atomic annihilation indicates one manner in which this aspect of technology as "black magic", the threat to man from his own instruments, has manifested.

As scientific technology now delves into areas that may impact human nature as such, like the human genome (creating possibilities for genetic modification) as well as possibilities for cybernetic interface between the human body and digital technology. This is before considering the possible direction of things like artificial intelligence about which Stephen Hawking warned:

> The genie is out of the bottle. We need to move forward on artificial intelligence development but we also need to be mindful of its very real dangers. I fear that AI may replace humans altogether.[20]

---

[17] Heidegger. The Question Concerning Technology, 312. (words in brackets are mine.)

[18] Heidegger. *The Spiegel Interview. (1966).*

[19] Berdyaev. *The Destiny of Man*, 227.

[20] Stephen Hawking. "I fear that AI May Replace Humans Altogether". An Interview with *Wired*.https://www.wired.co.uk/article/stephen-hawking-interview-alien-life-climate-change-donald-trump (Accessed. August 28, 2018).

This is why a contemporary theme is the idea of *transhumanism* –the process of replacement or "going beyond" the human species.. Heidegger and Hawking both captured the peculiarly modern anxiety that the processes of technology have slipped out of human control. This anxiety that grips much of modern life as evidenced in everything from environmentalism to dystopian science fiction, is that man's tools have escaped his grasp and have taken on a life of their own; that the ever uncontrolled and ever expanding power of the machine now threatens the human race either with outright annihilation, or with recasting man in its own image.

References

Aquinas, St. Thomas. 1897. Summa *Theologiae*. 1897. http://www.corpusthomisticum.org/sth3061.html#42226 Accessed May 2018.
Aristotle. *Nicomachean Ethics*. 1926. (1999 reprint). *Nicomachean Ethics*. Trans. H. Rackham, 1999. Loeb Classical Library, Harvard University Press.
——— Poetics. 1995. Longinus. On the Sublime. Demetrius. On Style. Trans. Stephen Halliwell (the Poetics.) Loeb Classical Library, Harvard University Press.
Condorcet, Marquis de. 1995. *Sketch for a Historical Picture of the Human Mind* in The Portable Enlightenment Reader. Isaac Kramnick(ed.) New York: Penguin.
Dawson, Christopher. "Catholicism and the Bourgeois Mind." At Catholic Culture.org. https://www.catholicculture.org/culture/library/view.cfm?recnum=2580. Accessed 7 Jan 2017.
Heidegger. 1966. Transcription of the Spiegel Interview "Only a God can save us" in http://www.ditext.com/heidegger/interview.html. Accessed 14 Aug 2018.
Hume, David.1987 (1742). *Essays, Moral, Political, and Literary*. Indianapolis, In: Liberty Fund. http://www.econlib.org/library/LFBooks/Hume/hmMPL24.html. Accessed May 2018.
Niebuhr, Reinhold. 1964. *The Nature and Destiny of Man. Volume I*. New York: Charles Scribner.
Rivington, James. 2013. 7 Scientific Breakthroughs that unlocked the Age of Technology. At TechRadar, https://www.techradar.com/news/world-of-tech/future-tech/physics-and-technology-1174316. Accessed May 2018.
Seneca, Lucius Annaeus. *Epistularum Moralem ad Lucillium* LXXXVIII (88) At the Latin Library. http://www.thelatinlibrary.com/sen/seneca.ep11-13.shtml. 5 May 2018.
Voltaire. 1924 (1995 translation.) *Philosophical Dictionary*. New York: Knopf. At Hanover. https://history.hanover.edu/texts/voltaire/volpreci.html. Accessed 4 Jan 2017.

Chapter 13
The Classical Ideal of High Culture in the Democratic Age

The rise of a commercial, technological civilization oriented toward the idea of progress into an ever improving future represented a repudiation of the veneration of antiquity. But perhaps nothing has rendered the classical ideal of culture more *morally* questionable to us than its points of tension with the principal moral and political value of modern age – that of *equality*. The Enlightenment culminated in what de Tocqueville termed the "Democratic Revolution." Thomas Paine's simple yet eloquent claim was that

> …all men being originally equals, no one by birth could have a right to set up his own family in perpetual preference to all others…[1]

This attack on the principle of aristocracy was no merely philosophical axiom, but the battle cry of the American and French revolutions which would eventually which would upend the old order of Europe and lead to the establishment of democratic ideals as the most energetic force in the political and cultural life of the modern West.

In contrast we must candidly admit that the classical ideal of culture had an aspect which was fundamentally "inegalitarian." It is true of course that democracy – the rule of the many – has Greek, indeed specifically Athenian origins in the work of great statesmen like Cleisthenes, Solon, and Pericles. Yet the Greek philosophers were singularly diffident about the value of democracy in both its political and cultural aspects. Socrates, the exemplar of philosophical virtue was executed by the Athenian democracy, casting doubt in the minds of his disciples on the wisdom and judgement of the multitude. In the *The Republic* as we know, Plato elaborates a theory of an ideal aristocracy where the educated and cultured philosopher-kings rule the others, and he treats democracy as one of the degenerate forms of constitution, given over to a chaotic licentiousness and treating unequal things equally.[2] As

---

[1] Thomas Paine. *Commonsense*. http://www.ushistory.org/paine/commonsense/sense3.htm (Accessed January 9, 2017).

[2] Plato.Republic VIII. viii *passim*.

for Aristotle while he accepts democracy("polity") where it aims at the common good as one of the legitimate forms of government, he warns pointedly that "... where the laws are not sovereign, then demagogues arise" and that "... a democracy of this nature is comparable to the tyrannical form of monarchy...[3]"

It is however, in the cultural sphere that the inegalitarian aspects of Hellenic civilization are most relevant to our present discussion. The Greeks believed passionately in *excellence* and strove above all to attain it. Greek culture was unsatisfied with the common and mediocre and took its motto from Homer – 'αἰὲν ἀριστεύειν"[4] – "ever to excel". This implied a fiercely competitive striving in which there will be difficult standards of accomplishment which few will attain. In accord with her deeply held principle, Greek education likewise exhibited a "canonizing" tendency which looked for works in the arts and letters –, such as Demosthenes in rhetoric, and Euripides in tragic poetry, and above all Homer which could serve as ideal standards for emulation and inspiration.[5] This approach of distinguishing the truly excellent from the mediocre was inherited by the Romans. Hence Quintilian in formulating his curriculum for the rhetor writes that he will choose *pauci enim <qui>sunt eminentissimi* "...a few authors – those who are the most eminent...".[6]

And such "classicism" century after century has continued to fertilize the European arts in much the same way. Works like the Parthenon in architecture, the Aphrodite of Milo or the Apollo Belvedere in sculpture, the drama of Sophocles, etc.... have set ideals for later artists and authors to emulate. And if European literary figured developed their own "national" classics in the modern languages – like Shakespeare in English, Dante in Italian, Cervantes in Spanish, etc. – it is remarkable how consistently these figures drew inspiration from the example of the ancients. And the same of course can be said of the visual arts (we need only consider the veneration of antiquity in the artists of Italian renaissance.)

So in spite of the Greek origins of democracy as a political form, arguably then the real source of this modern moral passion for human equality lies not in the classics but in Christianity with its conviction concerning universal human dignity in the sight of God and indeed its preferential concern for the poor, the weak, the suffering, and the marginal.[7] At its root the claim for human equality is not empirical but theological and moral. It is the conviction that all human beings are endowed by the Creator with an equal and infinite moral value and dignity. This conviction has proven morally fertile in Western history. The moral battles against slavery and racism, the mistreatment of women, and the exploitation of workers, as well as the enumeration of human rights all issue ultimately from this Christian spiritual foundation. It was the project of the Enlightenment to secularize the Christian

---

[3] Aristotle. *Politics* (1292ª).
[4] Homer. *Iliad.* VI. 206.
[5] Marrou. 161–163.
[6] Quintillian. 2001. *Institutio Oratoria/The Orator's Education.* Books 10.1, p. 274–275.
[7] The author is not in sympathy with the *critique* of the Christian ethic found for example in Nietzsche´s *On the Genealogy of Morals* (First Essay). However Nietzsche's genetic claim itself that Christianity introduced a kind of "underdog" morality with its ethos of mercy and compassion and concern for the poor, weak and oppressed is eminently defensible.

rooted ethical affirmation concerning universal and equal human dignity and convert it into a revolutionary *political* aspiration directed against fixed social inequalities.

It is not the task of this work to assay this democratic movement in the *political or moral* realm, the positive elements of which are indubitably clear to most readers. However, the issue of *cultural* democracy must enter into our topic for the claims of high culture are bound to be scrutinized in the context of triumphant egalitarianism because of its explicit claim to superior excellence. Indeed, the very charge of "elitism" is today often treated as a definitive argument which once made discredits the value of a cultural artifact. Allan Bloom in his work *The Closing of the American Mind* explained these democratic revulsions this way:

> Aristocracies hate and fear demagogues most of all, while democracies in their pure form hate and fear "elitists" most of all, because they are unjust, i.e. they do not accept the leading principle of justice in those regimes.[8]

And we must admit that there has been actual historical entanglement of high culture with formal and informal aristocratic elites. Why did this entanglement occur? Granting the cultural aristocratism of the classical philosophers, it is not immediately obvious what the relationship is between the classics and the actual historical nobility attacked by the modern democratic revolutions. To be born into a noble family is a good of fortune perhaps, but it is certainly no guarantee of superior intellectual gifts. Indeed, the argument of classical and renaissance philosophy moved in the direction of arguing that fitness for rulership rested on *vera nobilitas* (true nobility) which is acquired by education and virtue rather than by right of birth.[9]

For all that there *has* often been a peculiar alliance between the actual, historical aristocracies and the philosophers and artists. And as a historical matter, while the early European aristocracy was primarily a martial class, from the Renaissance onward the patronage not only of philosophers but of high culture generally was an increasingly important aspect of European aristocratic culture. To provide just a few examples Leonardo da Vinci found a patron Duke Ludovico Sforza, behind Mozart in the Prince-Archbishop of Salzburg, Haydn in the Esterhazy family, Shakespeare in the Earl of Southampton, and so forth. In short high culture benefited from its historical alliance with aristocratic patronage.

Why this might be the case is indicated by Allan Bloom in his discussion of the gentlemen as the historical allies of the philosophers:

> Why are the gentlemen more open [to philosophy] than the people? Because they have money, and hence leisure, and can appreciate the beautiful and the useless...Aristotle in his *Ethics* shows how the philosopher appears as the ally of the gentlemen...[10]

---

[8] Allan Bloom. *The Closing of the American Mind*. (New York: Simon & Schuster, 1987): 250. Herafter "Bloom".

[9] Cf. Quentin Skinner. *The Foundations of Modern Political Thought*. Vol 1 (Cambridge University Press, 2010): 236–238.

[10] Bloom, 279. (My brackets). This may derive also from Bloom's teacher Leo Strauss. See for example the discussion of the gentleman in "Liberal Education and Responsiblity" in *An Introduction to Political Philosophy: Ten Essays by Leo Strauss*. Hilail Gilden(ed.) ((Detroit,

In short, the gentlemen have in principle the leisure to appreciate those activities (as philosophy or the fine arts) which are truly noble, while those mired in the necessities of life are perforce obliged to focus on the merely useful.

In contrast to this form of high culture, mass culture means that the intellectual and aesthetic quality of culture is democratized, determined by the taste of the multitude, the people, the majority. For the classics this would be problematic. First, they would see the fundamental judgements concerning beauty and wisdom as involving issues of knowledge and truth rather than something which can be correctly determined by numerically adding up subjective opinions. Plato´s Socrates in the *Crito* criticizes reliance on the majority opinion – just as in matters of health one ought to follow the advice not of the majority but of the one who knows.[11] Of course if there is no truth in philosophy or in art then this argument fails – which perhaps explains the default relativism in much of modern aesthetic thought.

But if as the classics thought aesthetic, ethical, and intellectual judgement involves a form of knowledge– aesthetic taste for example must be *educated,* and the same may be said of moral judgements. But Aristotle for example was extremely pessimistic that the untutored and uneducated majority would seek anything beyond pleasure "The generality of mankind shows themselves to be utterly slavish, by choosing what is only a life for cattle."[12]

If the classics are correct then taking the majority view as the cultural arbiter and standard can exert a levelling, downward pressure.[13] In that case, popular taste and mass marketability will have a vulgarizing effect by re-orienting the culture around the most common aspirations of wealth, pleasure, and entertainment. This fact may help explain the increasing dominance of mass entertainment in popular culture and the corresponding declines of the venerable Western traditions in classical music, art, and philosophy. Necessarily, the standards of a mass market-based culture will have a deleterious effect on the more noble activities whose appreciation requires more thought, concentration, and education, and whose utility-value is less evident from a pragmatic perspective- such as philosophy, the liberal arts, and the fine arts. This problem was already identified by Nicholas Berdyaev in the 1930s:

> The mass determines what shall be the accepted culture, art, literature, philosophy, science, even religion. And there is no social demand for culture of a higher order, for spiritual culture, for real art or real philosophy. The social demand now is chiefly for technics, for applied natural science, for economics…[14]

We have seen already in Part I how the classical ideal of culture and the theoretic life in particular hinges on a kind of aristocratic sense of hierarchy among human activities. The mechanical or "banausic" arts aim at the merely necessary, while the

---

Michigan: Wayne State University Press, 1989): 323ff.

[11] Plato. Crito. (47a-48b).

[12] Aristotle. *Nicomachean Ethics* I.V.3.

[13] For an interesting if perhaps controversial discussion of the relation between equality and excellence see Charles Murray. *Human Accomplishment.*(New York: Perennial/Harper Collins, Copyright 2003): 450.

[14] Nicholas Berdyaev. *The Fate of Man in the Modern World.* (London, UK: Student Christian Movement Press,1935 reprinted n the USA): 112.

liberal arts (pace Seneca) are those proper to the free man who unburdened by necessity is free precisely to pursue the noble. Excellence (ἀρετή) according to Jaeger was "...the central ideal of all Greek culture".[15]

None of this is to argue of course against the real moral achievements which egalitarian movements have made which are inseparable from the Christian re-evaluation of the dignity of the manual laborer implied by the idea of God's Incarnation as a poor manual laborer. As Berdyaev wrote:

> The ancient Greco-Roman world despised work, did not consider it sacred and thought it only fit for slaves. That world was based upon the domination of aristocracy – democracy itself was aristocratic; and consequently the greatest philosophers of Greece, Plato and Aristotle, failed to see the evil and injustice of slavery...Christianity introduced a totally different attitude to labour. Respect for work and for workers is of Christian origin.[16]

Christianity with its carpenter God brought about over time a moral revolution in attitudes to workers to Western societies. And we may indeed be thankful that Christianity inculcated the idea of the dignity of labor and the laborer which – inherited by the Enlightenment – bore fruit in such achievements as the abolition of slavery, and the efforts to recognize the dignity of workers and improve their treatment. However, at times the egalitarian idea is carried to the point of denying the hierarchy of different forms of labor from whence it leads to anti-intellectualism and the philistine deprecation of philosophy and high culture. Berdyaev explains:

> ...culture implies a hierarchy of qualities, distinction between the quality of work, and personal gifts. Spiritual, intellectual and creative work is different in quality from physical labour which creates material goods, and it has a different place in the scale of values.[17]

While recognizing the equal human dignity of the intellectual and the manual laborer, one ought not to deny the hierarchical distinction between the activity of the body, and that of the mind or soul. We must not however consider this hierarchy as a Nietzschean embrace of "Master morality" with its contempt for the poor and weak.

The question before us rather is how to reconcile a *moral* egalitarianism which recognizes the equal worth and dignity of all human beings, with that degree of necessary *cultural* inegalitarianism required to sustain y standards of achievement in art, science, philosophy and all domains of culture. Turning to Berdyaev again:

> The Greco-Roman world left us a true conception of the qualitative value of aristocratically creative work, and it must be reconciled with the biblical and Christian idea of the holy and ascetic nature of labour and of the equality of all men before God.[18]

The "democratic" ethical concern evinced in Christian ethics for the vulnerable and weak and consequently for the manual laborer who may be exploited or otherwise treated with indignity, is different from, but not ultimately incompatible with

---

[15] *Op Cit. Paideia* I, 15.
[16] Nicholas Berdyaev. *The Destiny of Man*. (London: Geoffrey Bles, 1954): 215.
[17] Idem.
[18] Berdyaev. *Supra*. 215–216.

the "aristocratic" values of the Greeks with their aspiration for "the noble" manifested in the highest intellectual, aesthetic, and cultural achievements. A more valid question is how to sustain the ethical universalism which upholds the dignity of all and motivates the democratic culture of the modern age, while protecting the very existence of high culture against its levelling tendencies.

References

Aristotle. *Nicomachean Ethics*. 1926. (1999 reprint). *Nicomachean Ethics*. Trans. H. Rackham, 1999 Loeb Classical Library, Harvard University Press.
———. 1932. *Politics*. 1932 (2005 reprint). Trans. H. Rackham. Loeb Classical Library, Harvard University Press. Sometimes accessed at Tufts via Perseus –http://www.perseus.tufts.edu/hopper/text?doc=Perseus%3Atext%3A1999.01.0058%3Abook%3D3%3Asection%3D1280b. Accessed 22 May 16.
Berdyaev, Nicholas. 1935 (reprinted in the US.) *The Fate of Man in the Modern World*. London: Student Christian Movement Press.
———. 1954. *The Destiny of Man*. London: Geoffrey Bles.
Bloom, Allan. 1987. *The Closing of the American Mind*. New York: Simon&Schuster.
Homer. *Iliad*. VI. 206 Trans. Richard Lattimore. At http://homer.library.northwestern.edu/. Accessed May 2018.
Jaeger, Werner. 1962 (reprint). *Paideia: The Ideal of Greek Culture, Volume I*. New York: Oxford University Press.
Marrou, H. I. *A History of Education in Antiquity*. Trans. 1982 (1956 copyright). Trans. George Lamb. Madison: University of Wisconsin Press.
Murray, Charles. 2003. *Human Accomplishment*. New York: Perennial/Harper Collins.
Nietzsche, Friedrich. 1989 (reissue). *On the Genealogy of Morals and Ecce Homo*. Trans and Ed. Walter Kaufmann (and R.J. Hollingsdale).Vintage Press.
Paine, Thomas. *Commonsense*. At US History.org http://www.ushistory.org/paine/commonsense/sense3.htm. Accessed 9 Jan 2017.

Chapter 14
The Theoretic Life and the Challenge of American Pragmatism: Dewey and the Greeks in Contention

> *"I do most cordially hate you for writing against Latin Greek and Hebrew. I will never forgive you until you repent. No Never! It is impossible. –John Adams to Benjamin Rush. September 16, 1810 (https://founders.archives.gov/documents/ Adams/99-02-02-5563 (accessed May 5, 2018))*
>
> *"Were every greek & latin book (the New Testament excepted) consumed in a bonfire, the world would be the wise & better for it..."-Benjamin Rush to John Adams, October 2, 1810"*

One can say that the modern revolution that began in Europe culminates in America. The promise and dangers of modern commercial-technical civilization are found in their most acute highly developed form in America, which has become perhaps the most Baconian of nations. What has been the American project if not a great applied experiment in all the dominant ideas of modern thought? Where else do we find in greater degree or with greater success the advocacy of democracy, technological innovation, economic and political liberalism, and the valorization of commerce?

The rise of America to world power in the twentieth century, and its consequent pre-eminence in global political, economic, and cultural life are characteristic features of our time. The attention that America has given to wealth creation and technical power have made it the most powerful country on earth. No nation has more successfully harnessed the promise of modern technology into material wealth and power. America has led the world in technological revolutions which have wholly transformed the modern world -we need think only of its role of the airplane, Fordist methods of mass production, space travel, the internet and digital revolution.

The question we are bound to ask then is what is the space for the classical humanist ideal within this commercial-pragmatist cultural matrix? Historically, in fact there have been two strains in in American culture concerning the classical ideal of education was captured in the energetic correspondence between John Adams and Benjamin Rush. Rush fulminated against the classics as useless relics of a bygone aristocratic era fit at most for antiquarians and unsuited the modern age of democratic progress America represented. Society in short moved. On October 2, 1810 he writes to Adams:

> Were every greek & latin book (the New Testament excepted) consumed in a bonfire, the world would be the wise & better for it…a passion for what are called the Roman & greek Classicks may be compared to a passion for their Coins; they are well en'o to amuse the idle & the rich in their closets, but they should have no Currency in the modern pursuits and business of mankind.[1]

Adams replied on October 13 with equal vigor that the classics were the necessary foundation of cultural achievements:

> …I would put you into your own Transquilliser, till I cured you, of your Fanaticism against Greek an [SIC]Latin…. the World have never Seen a Milton if a Homer and Virgil had not lived before him…My Friend you will labour in vain. As the Love of Science and Taste for the fine Arts increases in the World, the Admiration of Greek and Roman Science and Litterature will increase.[2]

The American Constitution itself with its Senate and Republic reflects the immersion of the Founders in the classics. We have seen the Ciceronian John Adam's fervent advocacy for the classics. In *The Federalist Papers* whose authors took the name of "Publius" (from Publius Publicola one of the fathers of the Roman Republic) there are copious reference to the historical examples of Greece and Rome to settle points.[3] The Anti-Federalists were equally likely to take classical pseudonyms like "Brutus".

The classicism of the American founders however soon came into collision with the democratic-populist, commercial minded, and pragmatist strains in Americans which saw the classics as a relic of old world elitism. The rise of Jacksonian populism seems to be an early turning point. As Edward Miles wrote:

> In the first place, the traditional classical education was ill-adapted to the equalitarian-minded (if not equalitarian) Jacksonian era. Since the ability to read and quote Latin and Greek formerly served as a means of distinguishing the gentleman from the common man, such studies better suited a society that recognized rigid class distinctions. With their whole-hearted endorsement of democracy, many Americans frowned upon a system that tended to set men apart according to their presumed station in life.[4]

Yet together with that there has been an even stronger pragmatist strain which privileges practical achievements in commerce and technology over liberal learning. It is this critique of the liberal arts ideal from the side of American pragmatism which we must focus on as it is today one of the most formidable objections to the classical ideals of culture and education. From the standpoint of culture, the dangers that commercial and technical values will swamp those of classical humanistic culture are also naturally in evidence. What has principally made America a colossus

---

[1] https://founders.archives.gov/documents/Adams/99-02-02-5568 (accessed May 5, 2018).
[2] https://founders.archives.gov/documents/Adams/99-02-02-5570 (Accessed October 28, 2018)
[3] To pick just one of many examples see Madison's references to Solon, Draco, and Lycurgus in Federalist 38. http://avalon.law.yale.edu/18th_century/fed38.asp (accessed May 5, 2018).
[4] Edwin A. Miles. "The Young American Nation and the Classical World", 264.

in technology and commerce and a leader in global politics are the increasing predominance of the technical and commercial characteristics in American culture. These precise successes have augmented a pragmatist strain in American culture inhospitable to classical humanistic and intellectual ideals which privilege the theoretical over the practical.

We have seen that for the classics the exaltation of intellectual life for its own sake meant that manual labor and commerce as at best activities of a subordinate order, being merely necessary and useful. Yet in some manner American pragmatism tends to invert this hierarchy – its practical, technical and commercial orientation tending to exalt work, technology, and business, while tending to judge intellectual activity by its instrumental results in terms of profitability and usefulness.

This American focus on the pragmatic and commercial was critiqued in its earliest days even by America's friendliest foreign observers in ways we might today find excessive. We sometimes forget that this was a main element in de Tocqueville's very concept of American exceptionalism:

> The position of the Americans is therefore quite exceptional…Their strictly Puritanical origin, their exclusively commercial habits, even the country they inhabit, which seems to divert their minds from the pursuit of science, literature, and the art…have singularly concurred to fix the mind of the American upon purely practical objects.[5]

With respect to philosophy specifically Tocqueville comments that:

> I think that in no country in the civilized world is less attention paid to philosophy than in the United States. The Americans have no philosophical school of their own, and they care but little for all the schools into which Europe is divided, the very names of which are scarcely known to them…The Americans do not read the works of Descartes because their social condition deters them from speculative studies…[6]

Today of course these harsh judgements merit at the very least serious qualification. If American philosophers of the first rank have still been relatively scarce by comparison with Europe, America has since become a world leader in science both theoretical and applied; and in literature authors like Edgar Allen Poe, Herman Melville, Walt Whitman, and Mark Twain no doubt belong to any serious canon. The American university which we will discuss in the next chapter, long prized the classics and the humanities, and the American liberal arts college as well as the Great Books program of Mortimer Adler and Robert Hutchins have manifested a serious concerning America for the humanities. And America has produced original philosophy.

Still it is interesting that where American intellectual life has made its most original contributions to philosophy it is precisely in the philosophical school of *pragmatism*, which reflected in an intellectual register a basic American skepticism toward the classical intellectual ideal. Indeed, this polemic has become one of the

---

[5] Alexis de Tocqueville. Democracy in America. Book II. Chap. IX. http://xroads.virginia.edu/~hyper/detoc/ch1_09.htm (accessed 5/8/2017).

[6] Ibid. Book I.

most potent intellectual challenges to the theoretic ideal. This can perhaps be best studied by examining the principal exponent of the pragmatist school John Dewey. Dewey's objections can be classed under three broad rubrics each which also give voice to certain deeply rooted cultural tendencies in the broader American culture. These three are the themes of *progressivism, instrumentalism, and democratic-egalitarianism*. Let us take each before providing an evaluation and response.

14.1 Progressivism

We have already seen how Bacon and the Enlightenment inaugurated a new concept of history with their idea of progress. In Europe, however such modern progressivism has always had to contend with the consciousness of deeply rooted traditions. The violent upheavals that followed the French Revolution have no analogue in America. In America the modern ideals of the Enlightenment could be planted in a New World without the resistance of the Old. While other nations gloried in their past America saw itself (as John O'Sullivan stated it) the "great nation of futurity."[7] What this meant he fleshed out as follows:

> ...we are the nation of progress, of individual freedom, or universal enfranchisement...we must onward to the fulfillment of our mission – the entire development of our organization – freedom of conscience, freedom of person, freedom of trade and business pursuits, universality of freedom and equality.[8]

This progressivism penetrated deeply in to the American educational philosophy represented by John Dewey. Enamored of terms like "reactionary" and "antiquarian", Dewey reserves some of his strongest polemics against defenders of classical and Great Books education.

For Dewey there are three progressive developments revolutions which have opened an unbridgeable chasm between the modern world and the relatively primitive past. These are the scientific revolution, the industrial/technological revolution, and the democratic revolution. These case changes, which for Dewey are unqualified advancements, have rendered the older values of liberal education obsolete and inapplicable to the modern circumstance. Targeting the Great Books movement of Hutchins and Adler, Dewey makes clear its principal flaw is failing to recognize the superiority of the advancement which modern science and technology have brought over the philosophies of the past:

> The reactionary movement is dangerous (or would be if it made serious headway) because it ignores and in effect denies the principle of experimental inquiry and firsthand observation that is the life blood of the entire advance made in the sciences – an advance so marvel-

---

[7] John O'Sullivan. *Manifest Destiny*. https://www.mtholyoke.edu/acad/intrel/osulliva.htm (accessed January 30, 2017).

[8] Idem.

ous that the progress in knowledge in almost uncounted previous millenniums is almost nothing in comparison.⁹

At the core of his contention is that the great (and for Dewey benign) modern revolutions in science, technology, and democratization have essentially rendered the older tradition of the classics obsolete or of merely antiquarian interest. The Greek ideals of education and culture were historical products of *their* time not something of perennial value. While Dewey respects that Greek humanistic culture was well fitted to the social circumstances *of its day*, he regards those social arrangements as backwards relative to contemporary times. As he avers:

> This philosophy [Greek liberal education] was faithful to the facts of social life in which it appeared. It translated into intellectual terms the institutions, customs, and moral attitudes that flourished in the life of Athens…This fact might well make us look with suspicion upon an educational philosophy that, at the *present* day, defines liberal education in terms that are the opposite of what is genuinely liberal.¹⁰

14.2 Instrumentalism

Dewey's therefore engages in a wholesale rejection of the Aristotelian rooted distinctions we have discussed between the theoretical and practical arts, between the liberal and the mechanical, and between the intrinsically noble (such as truth) and the merely useful. At the core of Dewey's pragmatist theory is that truth IS nothing other than the useful and practically efficacious. He sees this indeed as something that follows from the scientific, experimental modes of knowledge of the modern era which in Baconian fashion are vindicated by their useful fruits:

> The hypothesis that works is the *true* one; and *truth* is the abstract noun applied to the collection of cases, actually foreseen, and desired, that receive confirmation in their works and consequences.¹¹

From this pragmatist perspective then the mechanical arts which aim at technical utility are precisely those which most meaningfully opened the knowledge of nature through their association with experimental science. Indeed, it was "…the marvelous advance of natural science…" which was the leading force in the:

> …breaking down of the wall existing in ancient and medieval institutions between "higher" things of a purely intellectual and "spiritual" nature and "lower" things of a "practical" and "material" nature.¹²

---

⁹ John Dewey "The Challenge to Liberal Thought" in *The Collected Works of John Dewey. Volume 15* (Carbondale, Southern Illinois University; Southern Illinois University Press, 2008); Dewey, 267 Hereafter*Works*.

¹⁰ *Works*. 262–263. (My brackets).

¹¹ John Dewey. *Reconstruction in Philosophy*.(New York, Henry Holt &Company, 1920, reprinted by Hardpress): 158.

¹² *Works. 271*

The modern *inversion* of the classical ideal today which deprecates the intellectual and the speculative in favor of the efficacious and profitable now receives in American pragmatism the imprimatur of philosophy itself. Needless to say, Dewey's conception has leaves room for the classical ideal of the theoretic life. The Platonic and Aristotelian conceptions of philosophy as the quest for a noble wisdom beyond all utility and profit are naught but the obsolete refuse of benighted, socially regressive, and pre-technological epochs in man's development. The good *is* the useful, and utility the ultimate test of truth.

14.3 The Democratic Ideal: Is Liberal Education Anti-Egalitarian?

America's revolt against the class distinctions of old Europe dividing the free gentleman from the manual laborer finds its correspondence in the plane of culture in that strain of American pragmatism which tends to devalue a "pure" theoretical intellectuality in favor of the value of such things as useful labor and mechanical invention. As Thomas Sowell explains a possible sociological analysis of the origins of this:

> From its colonial beginnings, American society was a "decapitated" society – largely lacking the topmost social layers of European society. The highest elites and the titled aristocracies had little reason to risk their lives crossing the Atlantic and then face the perils of pioneering. Most of the white population of colonial America arrived as indentured servants and the black population as slaves. Later waves of immigrants were disproportionately peasants and proletarians…the rise of American society to pre-eminence as an economic, military, and political power in the world was thus the triumph of the common man and a slap across the face to the presumptions of the arrogant, whether an elite or blood or books.[13]

Sowell here speculates that as a nation formed largely from Europe's laboring classes, a certain contempt for the unearned privileges of Europe's leisure class was carried to the new world. Indeed, he thinks the practical successes of Europe's underclasses in America served as a rebuke to the presumptions of old world elites. Logically, this disposition would tend to coincide with a rejection of the whole aristocratic ideal of culture which privileges intellectual and aesthetic contemplation over useful practical and manual labor. Thus, a certain resentment of those intellectuals given wholly to theoretical and speculative pursuits is intelligible given this historical association. With America's love of "the practical man" – from the frontiersmen to the modern entrepreneur – the most valorized forms of intelligence tend to be those devoted to useful, practical activities and technological invention (e.g. Thomas Edison.)

As we have seen part of what renders classical or liberal education, culture and modes of thought obsolete is not only the advances of modern science but the *social*

---

[13] Thomas Sowell. *The Quest for Common Justice.*(NewYork:Touchtone, 1999); 187.

progress toward a more egalitarian society, which we may call the democratic revolution.

For Dewey as we have seen the distinction between the liberal and mechanical arts is relic of pre-modern social conditions which sharply divided a servile class of manual laborers from free men. This division was not only undermined by the prodigious development of the mechanical or technological arts, but effectively overthrown by the democratic revolution in which America itself has played a vanguard role:

> At the time in which the scientific revolution was radically changing the nature and method of knowledge, understanding and learning, and in which the industrial revolution was breaking down once and for all the wall between the hand and the head, the political revolution of the rise of democracy was giving a socially free status to those who had been serfs. It thereby destroyed…the separation between "liberal" and "useful" arts.[14]

Here we see a somewhat ironic inner affinity between one of the foremost American intellectual schools of philosophy and the broader anti-intellectualism found in parts of American popular culture.

This is not of course to say that classicism and intellectualism found no place at all in America. We have briefly referenced the movement of perennialism also known as the Great Books movement led by figures like Robert Hutchins and Mortimer Adler in promoting the humanities and liberal arts education in American education. We might say that the Great Books approach to education was an effort to marry liberal ideals with those of American democracy. Indeed, Hutchins strongly promoted the universal extension of liberal education:

> The business of saying, in advance of serious effort that the people are not capable of achieving a good education is too strongly reminiscent of the opposition to every extension of democracy. The opposition has always rested on the allegation that the people were incapable of exercising intelligently the power they demanded.[15]

The effort to democratize liberal education by Adler and Hutchins were no doubt noble and also just in the sense that education ought not to be confined arbitrarily to a hereditary caste whose only qualification is accident of birth or fortune. Besides the inherent injustice, there is not least in such a system the accompanying loss of much untapped talent to society (though to a degree this view of things may be a polemical caricature of aristocratic cultures). All who have the talent and motivation ought obviously to have the opportunity to pursue liberal education should also have the opportunity.

Yet what remains unproven is whether the motivation to pursue liberal education is as democratically distributed as Adler and Hutchins supposed. This was precisely the skepticism of Plato who thought that liberal education:

> …would always tend to remain the privilege of an elite, since few were prepared to suffer the sacrifices it entailed and few could appreciate its advantages.[16]

---

[14] *Works*. "The Challenge of the Liberal Arts College", 277.

[15] Robert Hutchins. *The Great Conversation*. https://archive.org/stream/greatconversatio030336mbp/greatconversatio030336mbp_djvu.txt (Accessed 10/22/18)

[16] Marrou.38.

In a society like the United States in spite of a large middle class with ample leisure and opportunity, zeal for the humanities, for classical music and fine art, and for philosophy so far remains the province of a relatively few.

One must accept the possibility that in proportion as education has expanded democratically it has also been "massified" catering to more popular needs and desires met by a greater vocationalism in higher education.

At all events it was precisely the classical distinction between "liberal" and "mechanical" arts against which Dewey strongly rebelled in the name of democratic values. The un-democratic, class biased origin of the distinction is strongly emphasized:

> From Greece we inherit the tradition that puts "liberal" and "mechanical" in sharp opposition to each other…Vocational and practical education was illiberal in Greece because it was the training of a servile class. Liberal education was liberal in Greece because it was the way of life enjoyed by a small group who were free to devote themselves to higher things. They were free to do so because they lived upon the fruits of the labor of an industrially enslaved class.[17]

As America is for Dewey the land of democracy, in ought to leave behind the class ridden culture of Old Europe and the educational doctrines which followed upon its regressive, inegalitarian form of society.

> It is no accident that continental Europe, which is now the most disturbed portion of the world and the source of tragic disturbance everywhere else, is just that part that has stuck most closely to the educational philosophy we are now being urged to go back to. America must be looked upon either as an offshoot of Europe, or as a New World in other than a geographic sense.[18]

For Dewey this New World is called to realize the values of modernity – science, technology, vocationalism, and a democratic egalitarianism that will overturn the remnants of class hierarchy. To achieve this dream, it must:

> …banish the conception that the daily work and vocation of man are negligible in comparison to literary pursuits… It must accept wholeheartedly the scientific way, not merely of technology, but of life, in order to achieve the promise of modern democratic ideals.[19]

Perhaps Dewey's aims have succeeded beyond anything he could imagine. But instead of Dewey's hope that "*…the technical subjects we now call necessary acquire a humane direction*"[20] these relations have tended instead toward the wholesale swamping of the "liberal" and the "humane" by the "technical" and the "useful".

---

[17] *Works*. 262–263.
[18] *Works*, 274.
[19] Ibid. 275.
[20] Ibid. 279.

References

Adams, John and Benjamin Rush. Correspondence. At the Founder's Archives. https://founders.archives.gov/documents/Adams/99-02-02-5563 Accessed May 5, 2018, https://founders.archives.gov/documents/Adams/99-02-02-5568 Accessed May 5, 2018, https://founders.archives.gov/?q=John%20Adams%20%201810%20Rush&s=1111311111&sa=&r=24&sr Accessed May 5, 2018.

Dewey, John. 1920. *Reconstruction in Philosophy*. New York: Henry Holt &Company. reprinted by Hardpress.

———. 2008. *The Collected Works of John Dewey. Volume 15*. Carbondale: Southern Illinois University/Southern Illinois University Press.

Hutchins, Robert. *The Great Conversation*. https://archive.org/stream/greatconversatio030336mbp/greatconversatio030336mbp_djvu.txt. Accessed 22 Oct 2018.

Madison, James. *Federalist* 38. At Yale. http://avalon.law.yale.edu/18th_century/fed38.asp. Accessed 5 May 2018.

Miles, Edwin A. 1974. The Young American Nation and the Classical World. *Journal of the History of Ideas* 35 (2): 259–274. https://doi.org/10.2307/2708761.

O'Sullivan, John. *Manifest Destiny*. At Mr. Holyoke.edu https://www.mtholyoke.edu/acad/intrel/osulliva.htm. Accessed January 2017.

Sowell, Thomas. 1999. *The Quest for Common Justice*. New York: Touchtone.

Tocqueville, Alexis. *Democracy in America* At Virginia.edu. http://xroads.virginia.edu/~hyper/detoc/ch1_09.htm. Accessed May 2017

Chapter 15
A Reply to Dewey

15.1 Progressivism

We have seen that for Dewey, the classical argument in favor of the theoretic life is rendered obsolete by the reality of modern progress. The scientific and technological revolutions represent "an advance so marvelous that the progress in knowledge in almost uncounted previous millenniums is almost nothing in comparison."[1] The theory of progress – itself a fruit of Baconianism and the Enlightenment – poses the central challenge to any traditional view which venerates the classical ideals of the past. There are however a number of issues with this notion of progress. The first thing is that the scientific revolution itself cannot be set *against* the classical theoretical ideal, because it was built on it as its own foundation. It is true that the Greeks possessed no scientific-technological project comparable to what emerges in modernity. At the same time, what was absolutely necessary for this project was a product of the Greek theoretical mind. The modern scientific method brings together the Aristotelian principle of empirical induction with an advanced mathematics developed largely on the foundation of Greek mathematics (Euclid, Diophantus, etc....). This debt was, as we have seen, acknowledged by Aristotle's putative antitype Bacon when he wrote of Aristotle:

> But I, that should know best, do freely acknowledge, that I had my light from him; for where he gave me not matter to perfect, at the least he gave me occasion to invent."[2]

Nor is the relation between modern science and the classics a matter of purely antiquarian concern. It is instructive that late nineteenth century Germany considered the question of whether University education (particularly for students in the natural sciences) should be opened to those from the *Realschulen* which did not

---

[1] Dewey, 267.
[2] Robert Hutchins. "The Great Conversation" at http://blogs.britannica.com/wp content/pdf/The_Great_Conversation.pdf (accessed 5/19/2017). https://archive.org/stream/worksfrancisbaco02ba-coiala/worksfrancisbaco02bacoiala_djvu.txt – acces date 11/12/2015.

include a formation in the classics and in 1870 began to loosen the classical requirements. By 1880 the University faculty protested that students *in the physical sciences* with a Greek and Latin foundation did better than those whose formation is entirely modern i.e. science and technically based.

> It is also emphasized by the instructors in chemistry that graduates of *Realschulen* (Modern schools) do not stand upon the same level with graduates of *Gymnasia* (classical schools). Professor Hoffmann observes that the students from the *Realschulen,* in consequence of their being conversant with a large number of facts, outrank as a rule, those from the *Gymnasia* during the experimental exercises of the first half-year, but the situation is soon reversed, and given equal abilities, the latter almost invariably carry off the honours in the end; that the latter are better mentally trained, and have acquired in a higher degree the ability to understand and solve scientific problems.[3]

It is notable that this classically educated generation included so many of the figures who launched the last great revolution in modern physics. Heisenberg, Bohr, Planck, and Schrödinger themselves classically educated but the classics helped to inspire their creative achievements, with Heisenberg claiming that the study of Greek natural philosophy was essential to the development of modern atomic physics.[4] Thus it seems that the habits of mind and the development of creative intelligence instilled by the Greeks have helped to inspire the advance of the sciences down to last great efflorescence of physics in the late nineteenth and early twentieth century.

But there is an even more fundamental point beyond the *quid facti* which links the scientific revolution to the Greek theoretical achievements. No one can doubt that the knowledge of *facts* concerning the physical world have been vastly augmented by the modern sciences. The modern view for instance of the solar system, or the number and nature of the physical elements, or of the operation of physical forces on the changes in matter is far more accurate than the Aristotelian or Ptolemaic cosmologies and physics. In this restricted sense one may legitimately speak of progress.

But in the nobler questions which Greek philosophy wrestled with, the accumulation of empirical facts counts for little What is the good life for man? What is worth striving for? What is virtue? What is beauty? What is truth? It is folly to argue that say the manner in which the Aristotle's *Nicomachean Ethics* deals with the problem of virtue and the good life is somehow superannuated in the same way as his geocentrism. In short in all that concerns the highest things – wisdom and virtue and beauty and truth – the value of the Greek ideal of culture has proved perennial.

Progressivism has always assumed too close a connection between technological advances – which are obvious in the modern era – and progress in the intellectual and moral faculties of man. Going back to Condorcet:

---

[3] The quote comes from a report summing up faculty position by a Chemistry Professor A.W. Hoffman of Oct 15, 1880 cited in Richard Winn Livingstone. *A Defence of Classical Education.* (London, UK: Macmillan Press, 1917 – reprinted by *Forgotten Books*); 6–7.

[4] The claim is made by E. Christian Kopf. "Greek to Us: The Death of Classical Education and its Consequences". http://www.theimaginativeconservative.org/2012/03/greek-to-us-death-of-classical.html (Accessed March 8, 2017).

15.2 Democratic Culture and Instrumentalism

The sole foundation for belief in the natural sciences is this idea, that the general laws directing the phenomenon of the universe, known or unknown, are necessary and constant. Why should the principle be any less true for the development of the intellectual and moral faculties of man than for the other operations of nature?[5]

Yet the forward development of modern history with its vast destructive wars and concentration camps reveals the falsehood of Condorcet's premise – material progress is unrelated to moral progress. Technological advancement has indeed achieved the augmentation of human power. Considered in itself however, technology confers neither the wisdom nor the virtue to make beneficial use of it.

15.2 Democratic Culture and Instrumentalism

For a progressivist like Dewey the historical fact that the concept of the theoretical life and the liberal arts are bound up with an aristocratic ideal is in itself a decisive argument against their superior value. Since pre-modern societies were inegalitarian their social circumstances are inherently inferior to modern progressive democracy, so perforce must be their educational ideals.

The historical and social facts regarding the genesis of liberal arts tradition are beyond negation. Seneca as we have seen defined the liberal arts as those proper to "the free man" meaning one with sufficient independent wealth and hence leisure to devote himself to activities with their own excellence (rather than being absorbed in the merely necessary activities required for earning a living.)

This in itself seems to draw on Aristotle's analogy of philosophy with the free man; as the free man lives for himself and serves no other, so philosophy has its own value and serves no other science.

> ...for just as we call a man independent who exists for himself and not for another, so we call this the only independent science since it alone exists for itself.[6]

We must not shrink from the fact that the ideals of liberal education have always been more aristocratic than democratic in origin and tendency. It is doubtless true that the classical attitudes which exalted intellectual and aesthetic cultivation and subordinated commerce and manual labor was connected with notions of class and were inherited to a large degree by the European aristocracy and elites. And in passing we may ask: was this was wholly fruitless? It has become fashionable in our democratic age to see no redeeming merits at all in aristocratic culture; and indeed, its obvious inequities are easy enough to point out. Yet, one would want to take note of its positive contributions of the European nobility to the patronage of arts and letters. The aristocracy insofar as it served the important social role of being the protector of artists and thinkers helped bring to humanity imperishable achievements which all cultured persons can recognize. It is unlikely to be coincidental that

---
[5] In *Kramnick*, 26.
[6] Ibid. 982b.

the fortunes of European high culture have tended to wane as high culture lost aristocratic protection. As Berdyaev put it in a mass society where popular taste is the arbiter of culture:

> the intellectuals are socially defenceless; their existence is deprived of all material support. They are all too often compelled to feel their uselessness…[7]

At the same time is the liberal arts tradition really inextricable from a class structure such that it has no value independent of it? As Marrou puts the matter:

> It is undoubtedly true that the Greek's slave system allowed them to identify man – i.e. the "free" man – with the aristocratic man of leisure, who was relieved by the labour of others from performing any degrading work, and had every opportunity for indulging in a life of elegant leisure and spiritual freedom. But, I say again, the contingent forms of history are the bearers and embodiment of values which transcend them.[8]

While Aristotle presumed the social context of his own time, he acknowledges there is no one however leisured, who can devote themselves *entirely* to the higher rational activities of intellectual contemplation without also tending to the "lower" needs of the body. After all, since he affirms that "…in many respects human nature is servile…"[9] it is clear that for Aristotle the primary distinction is between liberal and illiberal *activities* not persons since all persons however favored by fortune are compelled by nature to engage in both kinds. Socrates, the pattern hero of philosophy was himself the poor son of a stone worker and midwife, Epictetus a slave. Thus, the liberal education tradition by no means required us to presume a rigid, impenetrable hierarchy of birth barring a "servile class" from the blessings of liberal education. In the Renaissance court school, figures like Vittorino da Feltre were famous for taking the poor as students provided they had the talent and will to pursue liberal studies.[10]

So, the really deep issue at stake is not the hierarchy of *persons* but the hierarchy of goods and ends and activities. We will recall Aristotle's statement that:

> Those things that are loved for the sake of some other thing (which one cannot live without) should be called "necessities" and secondary causes, but those that are loved for themselves, even if no other thing results from them, should be called goods in the strict sense…[11]

Hence the hierarchy that matters is the hierarchy of *sciences* and a*ctivities* with the practical and productive being subordinated to the theoretic activities.

Dewey's pragmatic concept of truth, and instrumentalism would seem to deny this distinction between noble goods and merely useful and instrumental ones. But

---

[7] Nicholas Berdyaev. *The Fate of Man in the Modern World*.(London, UK: Student Christian Movement Press,1935 reprinted n the USA): 112.

[8] Marrou. 222.

[9] Aristotle. *Metaphysics*.

[10] Paul Oskar Kristeller. "Humanism" in Minerva, Vol. 16, No. 4 (Winter, 1978): 587 and A,J. Clark "Renaissance Court Schools" in *The New Catholic Encyclopedia*. Volume 12. (Washington, D.C.: CUA Press, 1967): 369 and.

[11] From *Protrepticus*, B42 quoted A.W. Nightingale(ed.) *Spectacles of Truth in Classical Greek Philosophy,* (Cambridge University Press, 2004):194 (Greek texts removed).

are these ideas sustainable? The pragmatic conception which identifies truth with "what works" leads inevitably to contradictions. If one is forced to navigate by the stars, it makes more sense to think of the earth as a stationary point and the stars as orbiting around it. In other words, in this circumstance geocentrism "works". Yet for other applications thinking of the earth as a moving body is obviously more adequate. Are we to say that *both* heliocentrism and geocentrism are valid since both in some contexts "work"?

Manual labor is focused on the physical world and aims at transmuting the world of matter, while intellectual activity is essentially spiritual. But to deny for example that intellectual activity is higher than physical activity is in essence to deny the hierarchy which establishes the soul as higher than the body and reason as higher than the passions and appetites. The trajectory of Dewey leads not to an integration of these things, but to the supersession of the liberal by the illiberal, of the theoretical by the practical, of the noble by the utilitarian, of culture by technology. By the early twenty-first century, the classical ideal of the theoretic life has almost wholly given way to mass culture and the values of pragmatism, utilitarianism, and technical efficacy. The Baconian assault against the Aristotelian hierarchy of the sciences reaches in modern pragmatism its completion. In this lies the greatest threat to the classical ideal of theoretic life and its claim concerning the inherent nobility of wisdom and of the quest for it – philosophy.

References

Aristotle. *Metaphysics* 1933 (2003 reprint). Trans. Hugh Tredennick. Loeb Classical Library, Harvard University Press.

Berdyaev, Nicholas. 1935 (reprinted in the US.) *The Fate of Man in the Modern World*. London, UK: Student Christian Movement Press,1935 (reprint).

Clark, A.J. 1967. Renaissance Court Schools. In *The New Catholic Encyclopedia*, vol. 12, 369–370. Washington, D.C: CUA Press.

de Condorcet, Marquis. 1995. *Sketch for a Historical Picture of the Human Mind*. In *The Portable Enlightenment Reader*, ed. Isaac Kramnick. New York: Penguin.

Dewey, John. 2008. *The Collected Works of John Dewey*. Vol. 15. Carbondale: Southern Illinois University Press.

Hutchins, Robert. 1952. The Great Conversation. http://blogs.britannica.com/2008/12/the-great-conversation-robert-hutchinss-essay-for-the-great-books. Accessed May 2018.

Kopff, E. Christian. 2012. *Greek to Us: The Death of Classical Education and its Consequences*. http://www.theimaginativeconservative.org/2012/03/greek-to-us-death-of-classical.html. Accessed 8 Mar 2017.

Kristeller, Paul Oskar. 1978. "Humanism". Vol. 16, No. 4 (Winter, 1978.) Paul Oskar Kristeller. "Humanism" in Minerva. Vol. 16, No. 4 (Winter, 1978): 586–595. Springer (Publisher.). http://www.jstor.org/stable/pdf/41820353.pdf?refreqid=excelsior%3Aa8362333b0a7d243be0cfe9c8dffdf19. Accessed May 2018.

Livingstone, Richard Winn. 1917 *A Defence of Classical Education*. London, UK: Macmillan Press, 1917 – Reprinted by *Forgotten Books*.

Marrou, H. I *A History of Education in Antiquity*. Trans. 1982 (1956 copyright). Trans. George Lamb. Madison, WI: University of Wisconsin Press.

Nightingale, A.W. 2004. *Spectacles of Truth in Classical Greek Philosophy*. Cambridge: Cambridge University Press.

Chapter 16
The Contemporary Crisis of the Humanities: The Attack on the Western Canon and The Long Arm of Nietzsche, Marx, and Foucault

In a recent article for the *New York Times* Columnist Ross Douthat declares "Technocracy is Crushing the Life Out of Humanism." Citing data on the declining position of the humanities in American education he writes:

> Notably this trend is sharper among elite liberal arts colleges, the top thirty in the US News and World Report rankings, where in the early 2000s the humanities still attracted about a third of all students, but lately only get about a fifth. So, it's not just a matter of the post-Great Recession middle class seeking more practical degrees to make sure their students loans get repaid quickly; the slice of the American elite that's privileged enough and intellectually minded enough to choose Swarthmore or Haverford or Amherst over a state school or a research university is abandoning Hermes for Apollo at the fastest clip.[1]

The two factors seemed to be the augmentation of the trend toward technical-economic pragmatism combined with ideological assaults on the liberal arts tradition within humanities faculties themselves. If we confine ourselves only to the most acute contemporary manifestations of the "crisis of the humanities" within the English-speaking world, we must conclude the crisis is largely self-inflicted. By the late 1980s the traditional Western canon had become widely viewed by many professors and students within the humanities themselves as implicitly or explicitly racist, imperialist, and patriarchal, or at a minimal exclusionary toward women and people of color. Stanford University to take one example moved to remove its Western civilization core curriculum in 1988.[2] While there is nothing of course problematic about including the great works of non-European civilizations within a core curriculum, it is apparent that a self-loathing toward the civilization of the West and its classical texts has taken hold of large swaths of American academia. Efforts

---
[1] Ross Douthat. "Oh the Humanities", https://www.nytimes.com/2018/08/08/opinion/oh-the-humanities.html August 8, 2018 (Accessed August 29,2018).
[2] Richard Bernstein "In Dispute on Bias Stanford is Likely to Alter Western Culture Program."(Jan 9, 1988 -- https://www.nytimes.com/1988/01/19/us/in-dispute-on-bias-stanford-is-likely-to-alter-western-culture-program.html (accessed April 11, 2018)).

to re-introduce the Western civilization requirement in 2016 were met with angry rebukes:

> ...a Western Civ Requirement would necessitate that our education be centered on upholding white supremacy, capitalism, and colonialism, and all other oppressive systems that flow from Western civilizations....[3]

By the twenty-first century it had become possible to graduate from an Ivy League University from an English literature department without having read a single word of Shakespeare.

> The University of Pennsylvania does not require English majors to take an in-depth course on Shakespeare, which is also true at the vast majority of the country's most prestigious colleges and universities....[4]

The American university once served as an oasis for the liberal arts amid the broader societal trends toward pragmatism and utilitarianism. The Greek and Roman classics in fact were the dominant note of nineteenth century education. As J.H. Wright explained the *entrance* examination of Harvard University as of 1838 required a fluent knowledge of the works of Cicero and Virgil and the required undergraduate curriculum covered not only advanced Greek and Latin composition but the study of figures like Xenophon, Sophocles, Homer, Horace and Livy.[5] From the late nineteenth century onward however point the position of the classics and humanities has declined at a breakneck pace. Critics began to attack the system of classical education as part as of a broader critique of the past and the values inculcated by humane letters. As Paul Shorey wrote already in 1917:

> Greek and Latin have become mere symbols and pre-texts. They are as contemptuous of Dante, Shakespeare, Milton, Racine, Burke, John Stuart Mill, Alexander Hamilton, or Lowell, as of Homer, Sophocles, Virgil, or Horace. They will wipe the slate clean of everything that antedates Darwin's descent of man....[6]

To provide an example, by 1920 Greek was removed as a compulsory requirement for entrance into Oxford and Cambridge.[7] By the late 1950s Latin requirements also fell away.[8]

---

[3] Erika Kreeger. Op-Ed https://www.stanforddaily.com/2016/02/22/the-white-civs-burden/ (Accessed April 11, 2018).

[4] https://www.washingtonpost.com/news/answer-sheet/wp/2016/12/13/students-remove-shakespeare-portrait-from-english-department-at-ivy-league-school/?utm_term=.2e3c0c214bd7 (Accessed April 11, 2018).

[5] Wright, J. H. "Classical Education in the United States." *The Classical Review* 3, no. 1/2 (1889): 77–80. http://www.jstor.org/stable/690992.:78

[6] Paul Shorey. The Assault on Humanism. https://archive.org/stream/assaultonhumani01shorgoog/assaultonhumani01shorgoog_djvu.txt (Accessed June 13, 2018).

[7] Stray, Christopher A. "Culture and Discipline: Classics and Society in Victorian England." *International Journal of the Classical Tradition* 3, no. 1 (1996): 77–85. http://www.jstor.org/stable/30222253: 83.

[8] Ibid. 85.

Within America efforts to reinvigorate liberal education in the twentieth century came in the form of the Great Books approach, championed by figures like Robert Hutchins and Mortimer Adler. While conserving the core ideals of liberal education, this approach differed from more traditional classical education in two ways. First it consisted largely in reading books in translation rather than focusing on mastery of Greek and Latin philology. Secondly, it much further broadened the curriculum to include core texts of the modern world in science, literature, and philosophy. Finally, it took a democratizing approach aiming to introduce as many students as possible to the Great Books. It was this Great Books approach which was in large part the focus of critiques from the late 1980s onward. By the latter part of the twentieth century the health of the humanities as a whole seemed inversely proportional to the extension of higher education. According to statistics article by Peter Conn in *The Chronicle of Higher Education* the percent of humanities majors has dropped to just 8% of the student body in 2007, from 17.8% in the late 1960s.[9] While politicians frequently lament illiteracy in the sciences and mathematics, they seldom take note of the absence of basic literacy in humanities fields such as history. According to a study entitled *Still at Risk* conducted in 2008 by AEI researcher Frederick M. Hess, a majority of 17-year-old high school students could not place the US Civil War in the correct time period, nearly 40% could not identify the Renaissance, and nearly a quarter could not identify Adolf Hitler.[10]

The ultimate reasons for the self-immolation of the Western humanities, has its source in an intellectual critique, one ironically emanating not primarily from sources external to the European tradition but from two figures – Marx and Nietzsche – who themselves have a legitimate claim to be canonical figures in their own right. This critique has essentially two prongs. The prong deriving from Nietzsche we might say is *the suspicion of truth claims*. In his philosophical essay *On Truth and Lie in an Extra-Moral Sense,* Nietzsche writes:

> What then, is truth? A mobile army of metaphors, metonyms, and anthropomorphisms – in short, a sum of human relations which have been enhanced, transposed, and embellished poetically and rhetorically, and which after long use seem firm, canonical, and obligatory to a people: truths are illusions about which one has forgotten that this is what they are....[11]

Nietzsche here takes aim at the Socratic assumption underlying the whole of the Western tradition of the humanities, namely that the Truth exists and that its disinterested pursuit is inherently ennobling and gives meaning to the life of the mind.[12] The canonical texts in philosophy and literature were seen as expressions of this human quest for truth, goodness, and beauty, which Nietzsche declares to be chimerical.

---

[9] http://www.chronicle.com/article/We-Need-to-Acknowledge-the/64885 (accessed 5/3/2017).
[10] Frederick Heer. https://www.aei.org/publication/still-at-risk/ (accessed 5/3/17).
[11] Friedrich Nietzsche. "Truth and Lie in an Extra-moral Sense". http://oregonstate.edu/instruct/phl201/modules/Philosophers/Nietzsche/Truth_and_Lie_in_an_Extra-Moral_Sense.htm (Accessed April 11, 2018). I believe this is the Walter Kauffman translation.
[12] The antinomy of Socrates and Nietzsche on this and other points was an important theme for Leo Strauss and his pupil Allan Bloom.

The second prong of the attack derives from Marx. In his famous 1845 Theses on Feuerbach "The philosophers have only interpreted the world in varying ways, the point however is to change it."[13] In this Marx shows himself an heir of the basic pragmatist and utilitarian cast of modern thought. The intellectual must be an activist not a mere contemplator of reality on the Aristotelian model. And the intellectual's activity must be directed toward the liberation of the "underdog" from oppression by power and privilege.

Marx, like Nietzsche displayed his own form of suspicion toward truth claims rooted in the idea, that culture is a manifestation of the interests of the powerful. As Marx and Engels wrote in *The Communist Manifesto* "The ruling ideas of each age have ever been the ideas of the ruling class."[14] Hence for Marx it is vital to "…rescue education from the influence of the ruling class."[15]

From these two sources derive the pillars on which the critique of the Western canon is based. On the one hand the Truth does not exist; on the other hand, "truths" are an expression of the interests of the powerful and privileged.

The academic critique of the humanities traditions stems from new ideas concerning the role of the intellectual. One of the most influential of this new breed of intellectual, Michel Foucault stated in a published 1976 interview that:

> The important point here, I believe is that truth isn't outside power, or deprived of power… Each society has its regime of truth, its 'general politics' of truth…that status of those who are charged with saying what counts as true.[16]

What is called "truth" then is merely the reflection of shifting historical configurations of power. The role of the intellectual then is not so much to find the Truth, but rather to unmask the structures of power which lie behind the creation of "truths." An influential example of this idea can be found in the birth of post-Colonial. Edward Said's famous work *Orientalism* draws on Foucault's idea to reveal how Western scholarly representations of the Orient were in fact not disinterested scholarship but the ideological counterpart to European imperialism.

> …ideas, cultures, and histories cannot seriously be understood or studied without their force, or more precisely their configurations of power, also being studied. To believe that the Orient was created –or, as I call it "Orientalized" – and to believe that such things happen simply as a necessity of the imagination, is to be disingenuous. The relationship between Occident and Orient is a relationship of power, of domination….[17]

---

[13] Karl Marx. *Theses on Feuerbach.* https://msuweb.montclair.edu/~furrg/gned/marxtonf45.pdf (accessed 5/4/17).

[14] Karl Marx and Friedrich Engels. *The Communist Manifesto.* https://www.marxists.org/archive/marx/works/1848/communist-manifesto/ch02.htm (Accessed April 11, 2018).

[15] Idem.

[16] Michel Foucault. "Political Function of the Intellectual" https://www.radicalphilosophyarchive.com/wp-content/files_mf/rp17_article2_politicalfunctionofintellectual_foucault.pdf (Accessed 4/11/2018).

[17] Edward Said. *Orientalism* (Routledge&Keegan, 1978):13 in https://sites.evergreen.edu/politicalshakespeares/wp-content/uploads/sites/33/2014/12/Said:full.pdf. (accessed April 14, 2018).

16 The Contemporary Crisis of the Humanities: The Attack on the Western Canon...

The currently fashionable trends within the humanities sees the role of the intellectual as a political activist, unmasking and endeavoring to overthrow ostensible structures of privilege and power wherever they are found including behind the canon of the humanities themselves. In this they follow Marx's and Foucault's conception of the intellectual. The consequences of this with respect to the classics and more broadly the Western canon are plain enough. If ideas and truth claims are simply an expression of power configurations, then the canon cannot be a neutral repository of wisdom. Instead, it is the political expression of oppressive structures of race, class, and sex.

The classical human ideal of education and culture is at were caught in the pincers of ideologues of both "the right" and the "left".[18] The attack comes both from that part of "the right "which tends to reduces value to economic pragmatism and market utility, and from "the left" with its phobias of elitism and Eurocentrism deriving from its egalitarian imperatives is often hostile to the traditional Western canon. As Justin Stoyer eloquently put the matter:

> Vulgar conservative critiques of the humanities are usually given the greatest exposure, and yet at the same time, it is often political (and religious) conservatives who have labored the most mightily to foster traditional humanistic disciplines in the schools. Left defenders of the humanities have defended their value in the fact of an increasingly corporate and crudely economic world, and yet they have also worked to gut some of the core areas of humanistic enquiry— "Western civ and all that"—as indelibly tainted by patriarchy, racism, and colonialism.[19]

What may be said of such critiques? It is certainly also true that the great non-European civilizations like those of India, China, and the Islamic world have made important contributions to the collective human achievement which have not always received their due. Would it not a sign of a truly barbarous mind to be wholly devoid of admiration for the refined traditions of the Confucian classics in China, the brilliance of an Islamic philosopher like Avicenna, or the ethos of universal compassion to be found in the teaching of Goutama Buddha? Curiously however, the critics of the Western canon often seem far less interested in the positive task of understanding the religious, languages, histories, and classic texts of these non-Western cultures than in the negatively task of critiquing the real or supposed ills of Western civilization. If they were to explore the history and culture of the non-Western "Other" in depth, then in addition to a deeper appreciation of the achievements and positive attributes of these civilizations, they might also discover imperialism, war, slavery, and ethnocentrism are not exclusive properties of the West.

---

[18] The terms "left· and "right" are somewhat protean and arbitrary, but in the American context "the right" is often associated with classical liberalism – a support for the free market or capitalist system of political economy with minimal state interference, while the Left is association with egalitarianism particularly around issues centered on race, class, and sex.

[19] Justin Stover. "There is no Case for the Humanities"in *American Affairs*. (Winter 2017/Volume I, Number 4) https://americanaffairsjournal.org/2017/11/no-case-humanities/?utm_content=buffer23ea4&utm_medium=social&utm_source=facebook.com&utm_campaign=buffer. (accessed 1/3/2017).

Certainly, in the history of the West there have been evils and cruelties which merit moral condemnation. We may say much the same of other, non-Western civilizations. In both cases it would be as myopic to reduce a civilization to the evils which manifest in its history, as to ignore them.

Finally, it is simply an untenable fallacy to argue that the value of truth and beauty to be found in figures like Plato, Aristotle, Shakespeare or Galileo are somehow undermined by their personal sexual and racial identity. Except that is, on the premise that *there is no underlying truth* to be found, only the plays of power to be unmasked. But this premise reveals the underlying nihilism behind the attacks on liberal education. The great tragedy here is that unless the university can recover its reverence for the foundational texts of the civilization which spawned it, it cannot serve as an effective counter-weight against the mighty juggernaut of anti-intellectualism and economic pragmatism in the broader society which can find no use or value for the humanities at all.

References

Bernstein, Richard. 1988. In Dispute on Bias Stanford is Likely to Alter Western Culture Program. *New York Times.* https://www.nytimes.com/1988/01/19/us/in-dispute-on-bias-stanford-is-likely-to-alter-western-culture-program.html. Accessed 11 Apr 2018.

Conn, Peter. 2010. We Need to Acknowledge the Realities of Employment in the Humanities. *At Chronicle.* http://www.chronicle.com/article/We-Need-to-Acknowledge-the/64885. Accessed May 2018.

Douthat, Ross. 2018. *Oh the Humanities.* https://www.nytimes.com/2018/08/08/opinion/oh-the-humanities.html August 8, 2018. Accessed 29 Aug 2018.

Foucault, Michel. 1976. *Political Function of the Intellectual.* Published in *Politique Hebdo,* 247 from An interview in *Radical Philosophy,* 16. https://www.radicalphilosophyarchive.com/wp-content/files_mf/rp17_article2_politicalfunctionofintellectual_foucault.pdf. Accessed 4/11/2018.

Hess, Frederick M. 2008. Still at Risk. At AEI. https://www.aei.org/publication/still-at-risk/. Accessed 5/3/2017.

Kreeger, Erika. The White Civ's Burden. Op-Ed in *Stanford Daily.* https://www.stanforddaily.com/2016/02/22/the-white-civs-burden/ Accessed 11 Apr 2018.

Marx, Karl. 1976. *Theses on Feuerbach.* At Montclair.edu. https://msuweb.montclair.edu/~furrg/gned/marxtonf45.pdf. Accessed 5/4/2017.

Marx, Karl and Friedrich Engels. 1888. *The Communist Manifesto.* Trans. Samuel Moore (assisted by Engels.). https://www.marxists.org/archive/marx/works/1848/communist-manifesto/ch02.htm. Accessed 11 Apr 2018.

Nietzsche, Friedrich. *Truth and Lie in an Extra-Moral Sense.* Trans. Walter Kauffmann (Probable). At Oregon State. http://oregonstate.edu/instruct/phl201/modules/Philosophers/Nietzsche/Truth_and_Lie_in_an_Extra-Moral_Sense.htm. Accessed 11 Apr 2018.

———. 1989 (reissue). *On the Genealogy of Morals and Ecce* Homo. Trans and ed. Walter Kaufmann and R.J. Hollingsdale. Vintage Press.

Said, Edward. 1978 (2014). *Orientalism* London: Routledge & Keegan. At Evergreen: https://sites.evergreen.edu/politicalshakespeares/wp-content/uploads/sites/33/2014/12/Said:full.pdf. Accessed May 2018.

References

Shorey, Paul. Paul Shorey. The Assault on Humanism. The Atlantic Monthly Company, 1917. https://archive.org/stream/assaultonhumani01shorgoog/assaultonhumani01shorgoog_djvu.txt. Accessed 13 June 2018.

Stover, Justin. 2017. "There is no Case for the Humanities." In American *Affairs*. Winter 2017/ Volume I, Number. https://americanaffairsjournal.org/2017/11/no-case-humanities/?utm_content=buffer23ea4&utm_medium=social&utm_source=facebook.com&utm_campaign=buffer. Accessed May 2018.

Strauss, Valerie. 2016. "Students Remove Shakespeare Portrait From English Department at Ivy League School." Washington Post. https://www.washingtonpost.com/news/answer-sheet/wp/2016/12/13/students-remove-shakespeare-portrait-from-english-department-at-ivy-league-school/?utm_term=.2e3c0c214bd7. Accessed 11 Apr 2018.

Stray, Christopher A. 1996. "Culture and Discipline: Classics and Society in Victorian England." *International Journal of the Classical Tradition* 3, no. 1 (1996): 77-85. Accessed May 2017. http://www.jstor.org/stable/30222253: Accessed May 2018.

Wright, J.H. 1889. "Classical Education in the United States." in *The Classical Review* 3, no. 1/2 (1889): 77–80. http://www.jstor.org/stable/690992. Accessed May 2018, Accessed 5/3/2017.

Chapter 17
The New Protrepticus: A Concluding Exhortation to the Theoretic Life

> *We ought not to sail to the pillars of Hercules and run many dangers for the sake of wealth, while we spend neither labour nor money for wisdom* –Aristotle. Protrepticus (Aristotle. Protrepticus. Frg. 52. Jaeger. Aristotle. 59–60.)

Having surveyed the long trajectory of the Western intellectual tradition from its origins in classical Greek philosophy through the Baconian turn and forward to modern pragmatism, we are now in a position to make some assessments. We have seen that Western modernity as a distinctive world view begins with the peculiar genius of Sir. Francis Bacon, his critique of Aristotelianism and his re-conceptualization of the aim of knowledge and science as technological power. By now the fruits of this approach are clear and abundant.

Having been tested now for some four centuries, what may be said on behalf of Bacon's "modern idea" and the form of civilization it spawned? To its credit we can say that where once poverty was the almost universal norm, the modern techno-economic focus has uplifted the material conditions of broad masses of humanity. Human beings today in the most technologically developed nations on average live far longer. Infectious diseases like small pox and bubonic plague that once killed or maimed countless millions have been virtually wiped out. We have harnessed and mastered the forces of nature such that we can communicate almost immediately across the world and move across the globe with unprecedented speed. We have created consumer societies where middle class earners expect a wide availability of goods and constant improvements in the goods they purchase. With our machines we fly through the air like the birds, and even travel through space taking our first steps toward the stars. The Baconian turn has without doubt borne fruit beyond anything Bacon himself could have conceived of. So, given all these benefits on what grounds could one find fault in the turn away from the approach of classical philosophy with its antiquated cosmology and technologically fruitless concepts of nature?

17.1 Philosophy and the Good Life –Happiness and Virtue

We might first consider a question which deeply concerned the classical philosophers, which remains of greater and more universal human interest to humanity than the exploration of the cosmos - the problem of happiness and the good life. Unlike the ancient teleological cosmology, modern science is of little relevance to this question. Here all its laboratories and instruments are of no avail, while the classics represent a perennial font of wisdom and guidance. For the ancients since Socrates one of the great tasks of philosophy is precisely to seek knowledge concerning the good life or happiness (εὐδαιμονία).

To find happiness hinges more than anything else on choosing wisely among the possible goods in life one might chiefly pursue. What will make me happy? What is truly worth striving for? These are questions all thinking human these among the most fundamental questions all thinking human beings have. But how will one know which goods are most worthy of pursuit and most likely to lead to happiness without reflection upon them – i.e. without philosophy?

And the stakes of the question are high - for depending which goods one associates chiefly with happiness – money, pleasure, power, wisdom, health, etc.... one will live a different kind of life. And the consequences of choosing wrongly may well be unhappiness. One may for example judge that the pursuit of physical pleasures above all leads to happiness only to discover much like a drunkard awaking from a hang over the path of intemperate excess in physical pleasures leads to misery. In the busy, often frenetic pace of modern life this basic issue there often seems little time for adequate reflection. By default, one is perhaps most likely to unreflectively drift toward the common economic or hedonistic conceptions of "the good life". To "live well" is then thought to involve the achievement of great wealth and possessions which are thought to provide security and the ability to actualize one's desires. But what if one achieves wealth only to find it does not confer the happiness one sought? And what if one's setting of one's whole happiness on wealth and possessions becomes actually source of un-happiness when one finds oneself unable to bear their loss or lives constantly with the anxiety that changes of fortune may lead to their loss?

The classics as we have seen provided a clear answer "happiness [εὐδαιμονία] is activity in accordance with virtue"[ἀρετὴν].[1] The happy and flourishing life will be a life of intellectual and moral excellence. And since the intellectual and moral powers are both within and particular to man, such a life will be the most fully human. An understanding of happiness as something within the human being, provides a firm anchor and confers a certain tranquility of soul amid the inevitably shifting fortunes of life in which pleasures and pains, poverty and riches, health and sickness are often transitory and passing. As Epictetus put the matter:

---

[1] Aristotle. *Nicomachean Ethics*. X.vii.

17.1 Philosophy and the Good Life – Happiness and Virtue

> But what is philosophy? Does it not mean making preparation to meet the things that come upon us?[2]

What contributions to the good life does Baconian civilization offer? Chiefly its peculiar benefits lie in the realm of technology and economics – technical power and wealth. That both of these things may be *useful* to the good life when rightly employed is not in question. But can either qualify *as* the good life itself? In the Greek conception the supreme good would be one which is good in itself and so cannot be employed wrongly.

That the good life is not equivalent to technical power ought then to be evident. As the history of the twentieth century shows technical power has as many destructive as creative possibilities. The same technological power which made possible the aforementioned benefits also enables humans to destroy cities from the air, lay waste to whole continents, build totalitarian states, and to industrialize genocide. We may mention also the possibilities for a destructive relationship to the world of nature and other species which the modern environmental movements in particular give emphasis to. Technical power whatever useful benefits it may confer requires wisdom and virtue to guide it. It cannot in short be an intrinsic, final or ultimate good.

And what of the less ambiguously positive advancements technology has made possible? The advancements in medicine which have meant vast improvements in average life expectancy. Yet, even here there is room for relativization – a very Greek distinction was that between the question of living and *living well*.

Which of these concerns ought to occupy the first place in human striving was a question already clear in Homer with the dilemma of Achilles:

> I carry two sorts of destiny toward the day of my death. Either,
> If I stay here and fight beside the city of the Trojans,
> My return home is gone, but my glory shall be everlasting,
> but if I return home to the beloved land of my fathers,
> the excellence of my glory is gone, but there will be a long life.
> left for me, and my end in death will not come to me quickly.[3]

Broadly, speaking the mechanical arts and technology are concerned with the first issue - living itself. They aim at the maintenance and improvement of human life in its physical dimension and also in the accumulation of its material adornments. The liberal arts are concerned with the second, with living well more than quantitative longevity. And of course, the question "what is the good life?" is that which occupied much of classical thought from Socrates onwards. The point is not of course that longer life made possible by modern technology is no blessing; only that knowing how to live well is an even greater one.

It is here that they would no doubt relativize one of the true and proud achievements of the technological era – the advancements in medicine which have meant

---

[2] Epictetus. *Discourses* 3.10 (Oltather Trans.) quoted in Keith H. Seddon. *Epictetus* – at the Internet Encyclopedia of Philosophy. https://www.iep.utm.edu/epictetu/#SH4a (accessed May 2018).
[3] Homer. *Iliad*. 9:411–416 in http://homer.library.northwestern.edu/ (May 6, 2018) I used this Richard Lattimore translation removing the Greek text, the underlining, and the numeration.

vast improvements in average life expectancy. Yet, longevity by itself does not provide the wisdom to know how to *live* that life. A very Greek distinction between the question of living and *living well.*

And now we must turn to the other boon offered by the Baconian turn – the higher standards of living we experience. Granting the positive dimensions of improved material conditions, can the good life be considered *equivalent* to economic well-being?

17.2 The Tyranny of Utility: Pursuit of the Lower, Neglect of the Higher

Let us begin to approach this question with the greatest exemplar of philosophy as a way of life. We will recall that Socrates in his *Apology* made his plea for the value of his philosophical life by arguing with his statement that:

> I shall never give up philosophy or stop exhorting you and pointing out the truth to any one of you whom I may meet, saying in my accustomed way: "most excellent man, are you who are a citizen of Athens, the greatest of cities and the most famous for wisdom and power, not ashamed to care for acquisition of wealth, and for reputation and honour, when you neither care nor take thought for wisdom and truth and the perfection of your soul?[4]

It is true that Socrates said of himself that his wisdom was the knowledge of his ignorance. Yet he remained consistent in one central conviction: that the highest good is above all else *the good of the soul*. This is the center of his whole life and thought. If Socrates is correct then the quest for wisdom concerning the good of the soul – philosophy - will rank perforce among the foremost of human activities. And his exhortation was that men should look after their souls and seek the goods which perfect it – truth and virtue – with the same ardor and care they instead give to their bodies and possessions. As Jaeger says:

> This implies a Socratic hierarchy of values, and with it a new, clearly graduated theory of goods, which places spiritual goods highest, physical goods below them, and external goods like property in the lowest place.[5]

Hence according to Socrates one ought to care first for the soul, then for the body, and finally for external goods. This Socratic hierarchy of goods is inherited precisely by Aristotle:

> Now things good have been divided into three classes, external goods on the one hand, and goods of the soul and of the body on the other…those of the soul we commonly pronounce goods in the fullest sense and the highest degree.[6]

---

[4] Plato. *Apology.* 29d-e.
[5] Jaeger. *Paideia* II, 39.
[6] Aristotle. *Nicomachean Ethics*. I. Vii.23.

But this Socratic hierarchy is the direct contrary of the whole tendency of modern commercial and technological civilization. Modern civilization's argues correctly that technology is *useful*. But useful for what? Certainly not primarily for the "perfection of your souls" but primarily for *material* improvements either in human health (the good of the body) or even more commonly for the accumulation of "external goods." These things are goods insofar as they go, but goods of a lower order while neglecting goods of the higher order. Take for example the good of the body. It is doubtless true that human beings in developed nations are on average healthier and live longer. Yet as Aristotle said "…it is slavish to long for life instead of for *the good life*…"[7] Aristotle here pursues the Homeric and Socratic theme. Is it better to live a short life of nobility and excellence, or a long life of comfort?

An extreme example may suffice to amplify these questions. One of the persons during the *Titanic* tragedy who allegedly took the place of women or children on scarce lifeboats when the ship went down, lived longer than those they replaced. Yet it is easy to see that such longevity in shame is of less value than a short life of honor (think of those who chose to go down with the ship rather than behave ignobly.) It is the good life, not the mere fact of life which claims first importance. To place all the focus on mere living, (i.e. quantitative longevity) rather on questions like "what is life for? What is worth striving for? What does it mean to live well?" is really to make an idol of physical life, which whether long or short will inevitably end equally in death.

And what of the modern economic turn, with its focus on the production and accumulation of wealth and material goods which is an obsessive feature of modern consumer society. This even worse from the perspective of the classical philosophers. Man is superior to the goods of this world. To make external goods the center is to replace *human* value with the value placed on gadgetry and things. Hence to place one's focus on the acquisition of wealth is to serve lower things and neglect the higher good of man himself. Instead of the excellence of the soul we pursue large bank accounts, gadgets, comforts, pleasures, entertainments. What the classics would regard in relative terms as trifling things. We will recall what Aristotle said:

> …only the cultivated soul is to be called happy; and only the man who is such, not the man magnificently decorated with external goods, but is himself of no value. We do not call a bad horse valuable because it has a golden bit and costly harness; we reserve our praise for the horse that is in perfect condition.[8]

It is true that we can apprehend as goods things which pertain to the body – such as health and wealth and bodily pleasure. But we affirm these things as *unqualified* goods. As Socrates affirms it is only the *right* use of wealth and other external or bodily goods which can be called goods in an unqualified sense.[9] To for example use wealth for unjust ends is by definition contrary to the good. But to know what those just ends are pertains to the activity of the soul and not to the body. Hence the good

---

[7] Aristotle. *Protrepticus*. Frg. 52 quoted in Jaeger. *Aristotle*. Pg. 58 (my ital.).
[8] Ibid. Frg. 51, p. 57 in Jaeger.
[9] Plato. *Meno*. 78c-79a.

is properly speaking not the good of the body but the good of the soul. Such noble ends are the province of philosophy (in the classical sense), and not of technology which concerns itself with the sphere of the merely useful and instrumental.

17.3 Reversing the Socratic Turn: The Restriction of Reason

We will remember then that for Socrates the most important question is– "What is the good life?", and that Cicero credited Socrates's significance as turning the Greek philosophical mind to the problems of ethics. The study of nature for Socrates is of little value relative to question of the good life – the life virtue and happiness. As we may recall from Xenophon:

> He did not even discuss that topic so favoured by other talkers, "the Nature of the Universe": and avoided speculation on the so called "Cosmos" of the Professors, how it works, and on the laws that govern the phenomena of the heavens: indeed he would argue that to trouble one's mind with such problems is sheer folly.[10]

Modern scientific empiricism has in a sense reversed the Socratic turn which turned the attention of reason away from the nature of the universe and to problems of ethics and politics. We have seen how when Aristotelian teleology was rejected by modern philosophers, the bridge between the philosophy of nature and ethics was broken. Questions concerning the good and the good life are not within the purview of modern, empirically based science.

The empirical sciences *per definitionem* only deal with empirical things. They have proved masterful at divining the physical constitution of stars and planets, of the cells of animals, the atoms and molecules, and even the subatomic world and the fundamental forces and interactions which govern it. But the empirical sciences tell us virtually nothing about higher questions which Socrates approached like "what is the good life?" or "what is virtue?" or "what is beauty?" Questions concerning any kind of immaterial truth or reality in fact – God, the Good, the soul, freedom – stands entirely outside the purview of empirical science since these things are not realities perceived by the senses (even augmented by technological instrumentation.) A consequence of scientism is the tendency then to actually deny the existence of non-physical realities simply because they are not a proper domain of empirical science – thus begging the question of immaterial reality entirely. There is a tendency in modern culture see knowledge as empirical and scientific, while consigning fundamental questions of ethics and aesthetics to the merely subjective sphere. The relativity of values is therefore a natural conclusion of scientism.

The idea in short has developed that *only* scientific knowledge (in the sense of empirical knowledge) counts as knowledge, and consequently there is no possible knowledge of the good. But the new physical sciences structurally are incapable of

---

[10] Xenophon. *Memorabilia* 1.11. http://www.perseus.tufts.edu/hopper/text?doc=Perseus%3Atext%3A1999.01.0208%3Abook%3D1%3Achapter%3D1%3Asection%3D11 (Accessed 4/3/2015).

deal with the problem of the Good at all. And since the modern tendency tends to restrict reason to the empirical realm, problems of ethics tend to be seen as subjective rather than a realm in which we can speak of rational knowledge.

Even those who claim they can discern the origin of ethics in prehistoric evolution can only at most speak of certain positive facts in human evolutionary history. They cannot justify one systems of ethics as against another. Hence a consequence of the idea of the supersession of philosophy by science is the neglect of these higher and more important questions by reason.

The problems of natural theology, ethics and aesthetics excluded by restricting the domain of reason to empirical reality. And it is not only problems of ethics which are excluded by "scientism" but also broader issues of metaphysics. But even the premises of science – such as the validity of empiricism or the basic concepts – are essentially metaphysical (philosophical) and not questions which science itself can answer. This inability of the empirical sciences to adequately ground their own principles, methods, and assumptions in itself points to the need for a "meta-science" – i.e. a higher and more universal science.

17.4 The Ideal of Universal Science vs. "The Barbarism of Specialization"

Philosophy as the universal science is the proper nexus between the problems of physical sciences, mathematics, art, and religion – in short it belongs to philosophy to relate all the branches of knowledge to one another. The failure of the project of universal science means really *the dis-integration of the sciences*. Today there are only specializations. Such is what to be expected of a technically oriented age – specialists. But this is linked to what Ortega y Gasset incisively called "the barbarism of specialization":

> ...is only acquainted with one science, and even of that one only knows the small corner in which he is an active investigator. He even proclaims it as a virtue that he takes no cognisance of what lies outside the narrow territory specially cultivated by himself and gives the name of "dilletanttism" to any curiosity for the general scheme of knowledge.[11]

We may review these developments. Aristotle is the central figure in the initial emergence of European science. Within his schema there is universal science and there are the particular sciences. Within Aristotle's division of the sciences (e.g. Books Γ and E of the *Metaphysics*) physics or natural philosophy appears as a particular science within the universal science of philosophy. Physical science explores a particular *kind* of being – one that is changing and material (and we would naturally subdivide and particularize it further into the sciences which explore the living or the non-living, etc....). This universal science serves the function of providing a

---

[11] Jose Ortega y Gasset. *The Revolt of the Masses*. (New York/London: Norton, 1993 -reprint):110 (anonymous translator).

ground for the concepts of the particular sciences. For example, instance the ideas of the empirical scientific method, causality, time, space, life, matter, etc....are all concepts which require philosophical account or justification. Furthermore, the universal science will relate the particular sciences to the whole, by showing the interconnections of all the sciences within the universal system.

In treating physical science as a particular science, we presume there are other kinds of reality which can be explored by reason which are not sensible. Aristotle placed the mathematical sciences (which explore number, shape, and quantity) and natural theology (which explores God) among the theoretical sciences. Likewise, the problem of the good is dealt with in the practical sciences of ethics and politics, and so forth.

This possibility of the universal science thus hinges on the reality of the distinction between the intelligible and the sensible and the possibility of their unification in a science which embraces both. This works directly against the modern desire to reduce "science" to the sensible alone. The Enlightenment critique of metaphysics represented by Hume and Kant today passes for conventional wisdom. This is notable even in our linguistic tendency to restrict the meaning of "science" (from *scientia* -knowledge) to *physical* science.

Yet this move is in itself problematic since the foundations of science (in the modern sense) are not empirically grounded. The concepts with which science works – the ideas of being, cause, time, the validity of empirical knowledge, etc.... can only be known demonstrably by means of philosophical rationality not empirical verification. In this sense, and from this perspective, the perspective of knowledge rather than pragmatism, the need for "first philosophy" is no less today than in antiquity. Indeed, the vast augmentation of empirical knowledge by modern physics and the other natural sciences would seem to cry out for a new philosophical systematization of knowledge to bring unity to the project of the sciences.

Yet we will recall Hawking's claim that "philosophy is dead...scientists have become the bearers of the torch of discovery in our quest for knowledge." What would be the significance of such an event? To some degree it rests on an undeniable truth. What has occurred is that the particular sciences like physics, biology, astronomy, and chemistry and also mathematics in concert with them have all made remarkable progress over the last five centuries. Certain questions in Aristotelian physics at a high level of generality such as natural teleology or hylomorphism may still be more open questions than many will admit. Yet, though Aristotelianism provided a first historical foundation for the natural sciences, as a world system describing physical reality, it was clearly superseded by the Newtonian world system which in turn was superseded by the systems of Einstein and quantum mechanics.

Note however that while the particular sciences have moved forward, the universal science – philosophy – has given an impression of failure, and that by suicide more than destruction from without. Descartes's effort to re-ground "first philosophy" as a foundation for the particular sciences proved abortive under the powerful successive critiques of Hume, Kant, and Nietzsche. No real system which would integrate the new knowledge of the sciences has emerged. Philosophy has therefore seemed ever more uncertain and questioning of its own foundations. And given the

17.4 The Ideal of Universal Science vs. "The Barbarism of Specialization"

practical utility of empirical science in the sense of technology, philosophy has also increasingly been seen as irrelevant.

Dying then is the Renaissance ideal of *homo universalis* – "universal man" - the many-sided development of human powers and a mind acquainted with all the arts and sciences in their complex relations and inter-connections. The arts themselves in the Renaissance were deeply interconnected with the sciences, as we see in the use of mathematics to achieve visual perspective in painting, or the role of Leonardo's careful observations of nature in his art.[12]

Specialization is fitting for an age where wisdom is no longer seen to have intrinsic value and where technical education has superseded classical humanist education. Man is converted into a technician, a part of the machine, who only needs to understand his appointed function within the state or corporate institution. To reduce education to training for technical functions is to reduce man to his work. It is neither needed or wanted to know more than the task. A pure technicism treats man as a mere means to productive ends and in this subordination lies barbarism.

The hour is already late. Bacon dream of the "relief of man's estate" has been realized in the advanced material conditions of the modern age. At its best, this might have provided a broader leisure for many more to engage in the fruitful pursuit of intellectual and aesthetic enrichment. Instead, it seems the Baconian revolution's has culminated in a *technicism* which threatens to subordinate all higher culture to the vulgarizing claims of use-value, mass taste, and profitability. Today with vastly greater populations than in past times we find few if any figures of the caliber of Homer, Sophocles, Plato, or Aristotle or for that matter of St. Thomas Aquinas, Dante, Leonardo, Michelangelo, Shakespeare, Cervantes, Velasquez, Bach, or Mozart. Between the dominion of technological-economic values and an ideology of progress rooted in scorn for the inheritance of the past, our fine arts no less than our philosophy, liberal arts and humanities decline and wither. This means our age is threatened by nothing less than a new barbarism, a forgetfulness of the highest ideals of classical thought and culture which shined forth in the wisdom and accomplishment of the ancients. And the loss of their ideals means the neglect of the possibilities they offer for cultivating what is best and most noble in human nature itself. As Erasmus said centuries ago, so let us say today:

Sed in primis ad fontes ipsos properandum, id est graecos et antiquos.[13]

---

[12] See. Christopher Dawson. *The Dividing of Christendom* (San Francisco, Ignatius Press, 2008 reprint). 66.

[13] Erasmus. *De Ratione Studii*.III. *In* https://books.google.es/books?id=XXk8AAAAcAAJ&printsec=frontcover&source=gbs_ge_summary_r&cad=0#v=onepage&q=fontes&f=false (accessed June 21, 2018).

References

Aristotle. Nicomachean Ethics. 1926. (1999 reprint). *Nicomachean Ethics*. Trans. H. Rackham, 1999 Loeb Classical Library, Harvard University Press.

Dawson, Christopher. 2008 (Reprint). The Dividing of Christendom. San Francisco, Ignatius Press.

Erasmus, Desiderius. 1518. Argent, In https://books.google.es/books?id=XXk8AAAAcAAJ&printsec=frontcover&source=gbs_ge_summary_r&cad=0#v=onepage&q=fontes&f=false. Accessed 21 June 2018.

Homer. *Iliad*. Richard Lattimore Translation. At Northwestern. http://homer.library.northwestern.edu/. May 6, 2018.

Jaeger, Werner. 1962. *Aristotle: Fundamentals of the History of His Development*. Oxford: Oxford University Press.

———. 1963 (reprint). *Paideia: The Ideals of Greek* Culture. Vol II. (New York: Oxford University Press.

Ortega y Gasset, Jose. 1993 (Reprint.). *The Revolt of the Masses*. Trans. Anonymous. New York/London: Norton, -reprint.

Plato. 2001 (reprint). *Euthyphro. Apology. Crito.Phaedo*. Phaedrus. Ed. Jeffrey Henderson. Loeb Classical Library, Harvard University Press.

Seddon, Keith H. *Epictetus*. Internet Encyclopedia of Philosophy. https://www.iep.utm.edu/epictetu/#SH4a. Accessed May 2018.

Xenophon. 1923 (2013 revision). Memorabilia. Oeconomicus. Symposium. Apology. Trans. E.C. Marchant. O.J. Todd. Loeb Classical Library, Harvard University Press.

Xenophon. *Memorabilia*. At Perseus: http://www.perseus.tufts.edu/hopper/text?doc=Perseus%3Atext%3A1999.01.0208%3Abook%3D1%3Achapter%3D1%3Asection%3D11. Accessed 4/3/2015.

Index

A
Adams, J., 129, 130
Adler, M., 131, 132, 135, 147
Alberti, L.B., 20
America, 3, 18, 129–132, 134–136, 147
Aristocracy, 22, 117, 123, 125, 127, 134, 141
Aristotle, 6, 12, 29, 35, 37, 49, 63, 71, 77, 81, 96, 97, 116, 124, 139, 150
Augustine, 15, 85, 114

B
Bacon, Sir F., 52, 61, 65, 71–75, 78, 79, 82–84, 87, 88, 93–95, 97, 100, 105, 114, 121, 132, 139, 153, 161
Barbarism, 3–9, 114, 159–161
Being, 4–6, 8, 13–15, 25, 34, 38, 40–42, 45, 51, 55, 56, 64, 77, 79, 81, 82, 85, 88, 90, 102–104, 106, 116, 141, 159, 160
    See also Ontology
Berdyaev, N., 28, 114, 120, 121, 126, 127, 142
Bloom, A., 3, 125, 147

C
Capitalism, 117, 146
Causality, 39, 79, 81–88, 101, 102, 104, 105, 160
Cicero, M.T., 3, 7, 15, 19, 21, 23, 35, 39, 146, 158
Classics, 3, 17, 71, 115, 117, 124–126, 129–131, 133, 139, 140, 146, 149, 154, 157
Commerce, 22, 24, 75, 77, 114, 117, 129–131, 141

Culture, 3–9, 20, 22–24, 27, 28, 33, 62, 65, 66, 74, 77, 96, 114, 118, 119, 123, 129–136, 140–143, 145, 146, 148, 149, 158, 161

D
da Vinci, L., 21, 125
de Condorcet, M., 114, 115, 140, 141
de D'Álembert, J., 94, 95
Deduction, 53
de Maistre, C.J., 93
Democracy, 33, 96, 123–125, 127, 129–131, 135, 136, 141
Descartes, R., 56, 62, 63, 69, 74, 79, 82, 84–88, 95, 98, 103, 107, 121, 131, 160
de Tocqueville, A., 123, 131
Dewey, J., 133, 139–143
Diderot, D., 32, 94, 95
Digital age, 119

E
Economics, 3–9, 14, 18, 24, 29, 63, 64, 117–119, 126, 129, 134, 145, 149, 150, 153–157, 161
Education, 3, 4, 6–8, 12, 17–20, 24, 25, 33, 65, 119, 124–126, 129, 130, 132–134, 139–142, 145–150, 161
Egalitarianism, 96, 114, 125, 127, 132, 136, 149
Einstein, A., 108, 109, 119, 160
Empiricism, 84, 85, 95, 96, 99–101, 158, 159
Encyclopedia, 6, 17, 91, 94, 95, 102, 142, 155
Epistemology, 100, 102

Erasmus, 3, 161
Ethics, 8, 13–15, 29, 31, 37, 38, 41–47, 50, 54, 57, 62, 66, 69, 79, 81, 82, 87, 89, 90, 116, 125–127, 140, 154, 156, 158–160
Europe, 3, 7, 18, 61, 79, 94, 123, 129, 131, 132, 134, 136

F
Faustianism, 28
Final causality, 39, 79, 81–84, 86–88
First philosophy, 16, 17, 69, 75, 84, 86, 96–99, 160
Foucault, M., 145–150

G
Galileo, 86, 94, 150
Gellius, A., 24
God, 5, 31, 55, 85–88, 104, 105, 124, 127
The good life, 29–31, 37–47, 69, 89–91, 96, 140, 154–158
Great Books, 4, 131, 132, 135, 147

H
Happiness, 25, 41, 42, 50, 98, 154–156, 158
Hawking, S., 97, 99, 121, 122, 160
Hegel, G.W.F., 97, 106
Heidegger, M., 28, 64, 120–122
Hobbes, T., 43, 89–91
Homer, 18, 19, 42, 46, 51, 124, 130, 146, 155, 161
Hugh of St. Victor, 22, 77
Humanism, 3, 7, 19, 20, 24–25, 35, 73, 114, 121, 122, 142, 145, 146
Humanitas, 3, 24, 25
Hume, D., 93, 97–110, 117, 160
Husserl, E., 4
Hutchins, R., 131, 132, 135, 139, 147

I
Induction, 52, 68, 139
Instrumentalism, 132, 133, 141–143

K
Kant, I., 93, 95, 97–110, 160
Kepler, 86, 94
Knowledge, 5–8, 13–18, 24, 27–35, 37, 43, 49, 50, 52–54, 56, 57, 62, 63, 65–69, 71–75, 78, 79, 81–85, 87, 90, 93, 94, 97–100, 102–105, 114, 115, 118, 119, 126, 133, 135, 139, 140, 142, 146, 147, 153, 154, 156, 158–160
See also Epistemology

L
Liberal arts, 5, 6, 9, 12–25, 64, 77, 116–118, 126, 127, 130, 131, 135, 141, 142, 145, 146, 155, 161
Life of enjoyment, 42–45
Logic, 19, 20, 52, 66, 118

M
Machine civilization, 113
Marx, K., 4, 63, 119, 145–150
Mathematics, 4, 19, 33, 49, 55, 65, 77, 84, 86, 94, 99, 139, 147, 159–161
Maxwell, J.C., 107, 119
Mechanical arts, 5, 9, 14, 21–24, 75, 77–79, 116, 133, 135, 155
Medieval, 5, 9, 19–22, 61, 63, 65, 67, 68, 72, 78, 79, 88, 98, 117, 133
Metaphysics, 6, 12, 14, 16, 35, 38, 39, 53, 66, 69, 71, 74, 75, 77, 82–85, 89, 96–110, 159, 160
See also First Philosophy
Michelangelo, 21, 161
Minkowski, H., 109
Modernity, 4–6, 9, 28, 30, 38, 56, 61, 62, 64–66, 74, 81, 96, 136, 139, 153
Music, 5, 19, 20, 22, 126, 136

N
Nasr, S.H., 61, 84
Nature, 8, 28, 38, 51, 62, 71, 77, 81, 94, 102, 115, 124, 133, 140, 153
Newton, Sir I., 86, 94, 105–110
Niebuhr, R., 114, 115
Nietzsche, F., 27–29, 64, 124, 145–150, 160
Noble, 5, 6, 9, 12–25, 42, 44–46, 50, 51, 55, 71, 74, 77, 116–118, 125–128, 133–135, 142, 143, 158, 161

O
Ontology, 38, 104

P
Philosophy, 4, 12, 29, 35, 37, 51, 61, 72, 77, 81, 93, 97, 114, 125, 131, 140, 147, 153

Index 165

Physics, 8, 38, 39, 55, 65, 79, 82–84, 86, 88, 89, 94, 105–110, 118, 140, 159, 160
Plato, 8, 13, 19, 20, 22, 23, 29–35, 37, 39–44, 49, 63, 65, 74, 78, 79, 82, 83, 85, 88, 99, 114, 117, 123, 126, 127, 135, 150, 156, 157, 161
Pleasure, 20, 21, 34, 42–45, 47, 51, 79, 81, 90, 126, 154, 157
Poetry, 5, 12, 18–20, 33, 124
Political life, 19, 29, 30, 35, 37, 38, 42, 46–47, 49–51, 77, 89, 91
Politics, 8, 29–31, 33–35, 38, 40, 43, 46, 50, 51, 62, 79, 81, 82, 89, 91, 98, 124, 131, 148, 158, 160
Power, 4–7, 19, 24, 25, 28, 30, 31, 33–35, 43, 45, 47, 62, 63, 66, 67, 71–74, 78–80, 82, 84–86, 91, 94, 100, 101, 110, 114, 115, 118–122, 129, 134, 135, 141, 148–150, 153–156, 161
Pragmatism, 3–9, 118, 143, 145, 146, 149, 150, 153, 160
Progress, 31, 66, 67, 78, 83, 86, 89, 106, 113, 119, 120, 123, 130, 132, 133, 135, 139–141, 160, 161
Progressivism, 66, 113, 132, 139–141

R
Rationalism, 27, 28, 30, 32, 84, 95, 100, 104
Relativity, 107–109, 158
Renaissance, 3, 7, 18–21, 25, 35, 61, 64, 65, 73, 114, 118, 124, 125, 142, 147, 161
Rhetoric, 19–21, 33–35, 44, 63, 124
Roman Catholicism, 66
Romans, 3, 15, 18, 19, 24, 124
Rush, B., 129

S
Sadoleto, J. (Cardinal), 7, 8, 25, 65
Said, E., 148
Science, 5, 14, 27, 50, 62, 71, 77, 82, 93, 97, 114, 126, 130, 139, 147, 153
Secularism, 61, 62
Seneca, L.A., 15, 18, 20, 23, 77, 116, 117, 127, 141
Smith, A., 4, 117
Socrates, 8, 13, 18, 19, 22, 27–35, 37–39, 41, 44, 53, 63, 72, 89, 123, 126, 142, 154–158
Soul, 5–7, 18, 23, 25, 28–30, 34, 40, 42, 44–46, 74, 79–81, 90, 101, 104, 127, 143, 154, 156–158

Space, 103, 106–109, 129, 153, 160
Spinoza, B., 69, 79, 82, 87
Strauss, L., 30, 31, 62, 81, 82, 88, 90, 91, 95, 96, 105, 125, 147
The supreme good, 13, 40–45, 69, 74, 89–91, 116, 155

T
Technology, 4, 6, 7, 22, 28, 62–64, 71, 75, 77, 81, 94, 99, 114, 117, 119–122, 129–132, 136, 141, 143, 155, 157, 158, 161
Teleology, 37–47, 69, 79, 81, 158, 160
Time, 3, 6, 7, 17, 18, 21, 22, 25, 30, 34, 35, 41, 42, 54, 61–69, 77–79, 94, 97, 98, 102, 103, 105–109, 114, 115, 117, 119, 127, 129, 133, 135, 139, 142, 145, 147, 149, 154, 160, 161
Theoretic life, 6–9, 21, 27–35, 37, 42, 44, 46, 47, 49, 65, 67, 69, 74, 75, 77–79, 81, 82, 89, 90, 92, 96, 97, 99, 105, 126, 129, 153–161
Theoretical science, 52, 54, 55, 77, 99, 105, 160
Thomas Aquinas, St., 5, 15, 41, 55–57, 66, 78, 98, 100, 116, 161

U
University, 3, 7, 8, 12, 13, 19, 20, 25, 27, 29–33, 35, 65, 66, 72, 81, 91, 93, 95, 97, 98, 103, 125, 126, 131, 133, 139, 140, 142, 145, 146, 150
Utility, 4–6, 8, 12–14, 16, 17, 21–23, 42, 43, 46, 53, 54, 64, 65, 68, 69, 73–75, 77, 78, 84, 97, 99, 115, 116, 118, 126, 133, 134, 149, 156–158, 161

V
Virtue, 6, 8, 19, 20, 25, 29–31, 33–35, 37, 40, 42, 45–47, 49–51, 56, 62, 63, 74, 79, 81, 91, 118, 125, 140, 141, 154–156, 158, 159
Visual arts, 124
Voltaire, 32, 94, 114, 115

W
Western Canon, 4, 145–150

GPSR Compliance

The European Union's (EU) General Product Safety Regulation (GPSR) is a set of rules that requires consumer products to be safe and our obligations to ensure this.

If you have any concerns about our products, you can contact us on

ProductSafety@springernature.com

In case Publisher is established outside the EU, the EU authorized representative is:

Springer Nature Customer Service Center GmbH
Europaplatz 3
69115 Heidelberg, Germany